CAMBRIDGE COUNTY GEOGRAPHIES

SCOTLAND

General Editor: W. Murison, M.A.

T0352328

DUMBARTONSHIRE

Cambridge County Geographies

DUMBARTONSHIRE

by

F. MORT,

D.Sc., M.A., F.G.S., F.R.S.G.S.

With Maps, Diagrams and Illustrations

CAMBRIDGE
AT THE UNIVERSITY PRESS
1920

CAMBRIDGE UNIVERSITY PRESS
Cambridge, New York, Melbourne, Madrid, Cape Town,
Singapore, São Paulo, Delhi, Mexico City

Cambridge University Press
The Edinburgh Building, Cambridge CB2 8RU, UK

Published in the United States of America by Cambridge University Press, New York

www.cambridge.org
Information on this title: www.cambridge.org/9781107678774

First published 1920
First paperback edition 2013

A catalogue record for this publication is available from the British Library

ISBN 978-1-107-67877-4 Paperback

CONTENTS

ILLUSTRATIONS

MAPS

Note. For the illustrations on pp. 5, 27, 30, 32, 40, 41, 64, 74 the writer is indebted to J. W. Reoch, Esq.; those on pp. 38, 42, 43, 72 are from photographs by the author; that on p. 92 is reproduced by permission of the Cunard Steamship Co. Ltd.; those on pp. 95, 97, 99 by permission, from photographs kindly supplied by Messrs William Beardmore and Co., Ltd.; those on pp. 114, 116, 117 are reproduced from Dr G. Macdonald's *The Roman Wall in Scotland* by arrangement with Messrs Jas. MacLehose and Sons; that on p. 140 by arrangement with Messrs T. and R. Annan and Sons; that on p. 149 by permission of *The Scottish Field*; and those on pp. 6, 7, 10, 19, 23, 26, 36, 53, 56, 58, 60, 62, 65, 94, 104, 124, 125, 143, 144, 147, 148, and 150 are from photographs supplied by Messrs Valentine and Sons, Ltd.

1. County and Shire. The Origin of Dumbartonshire.

The modern county of Dumbarton represents the greater part of the old district of The Lennox, which embraced much of southern Stirlingshire and also part of Perth and Renfrew. The Earldom of Levenach—corrupted into Lennox—was created by William the Lyon in the twelfth century, and given to his brother David, Earl of Huntingdon. Thus the hero of Scott's *Talisman* was the first recognised earl of the county. Most of the other counties of Scotland have had a similar origin. They were originally earldoms governed by the king's friends, who in many cases took their titles from the districts that they ruled. When William I had conquered England, large tracts of land were given to his companions or *comites*. Each district was therefore called a *comté*, from which we get the word *county*. The word *shire* is of Anglo-Saxon origin, and is derived from *scir*, administration, province.

The modern county is a political unit. As a rule we think of it as the division administered by a county council, but from the very beginning it has been the part of the kingdom under the jurisdiction of a sheriff. Originally the sheriffdom was a hereditary office held by the chief nobleman of the district. This early arrangement tended to instability. It was to the advantage of the noble sheriff to increase his territory. But his rival earls, the king, the church, and the burghs were also struggling to increase their power at other people's expense. Thus in early times the county boundaries were subject to frequent alterations. One factor, however, was permanent, and exerted a steady influence always in one direction. The dominant physical features of a country tend always to mould the political divisions in harmony with natural regions, and the boundaries of most Scottish counties show some adjustment to important physical features. Dumbarton is not so good an illustration of this principle as the neighbouring counties of Renfrew and Lanark, yet part of its boundaries have a well-defined geographical basis. Lanarkshire is a compact, geographical unit, consisting simply of the upper and the middle Clyde basin. The point at which a large river becomes too wide to be bridged conveniently is of prime importance. The stream of traffic down the valley divides here, and the up-river traffic coalesces at this point. Near such a point, counties often terminate; and this is the case with the Clyde. Above this point the banks of the river are embraced by one county. Below it, the river forms the boundary between Renfrew and Dumbarton. Most of the western boundary of Dumbarton is formed by the striking natural barrier to communication interposed by Loch Long, which for its whole length separates Dumbarton from the ancient district of Cowal. The north-east boundary is also a natural one. The dividing line runs through Loch Lomond from head to foot. The boundary of the south-eastern part of the shire is a historical compromise. Dumbarton here touches Stirling, Lanark, and Renfrew, and frequent readjustment of the boundary took place in accordance with purely historical factors. The origin of the detached part of Dumbartonshire will be considered in a later chapter.

Parts were lopped off the ancient district of the Lennox from time to time, until by the thirteenth century it coincided practically with the present county of Dumbarton. About this date the old name of Lennox was given up and the modern name adopted In the reign of David II there was an important readjustment of boundaries, and finally at the beginning of the sixteenth century the present limits were fixed with one small exception. In 1891, the Boundary Commissioners increased the area of Dumbarton by making the boundary include the town of Milngavie and all the parish of New Kilpatrick, which hitherto had been partly in Stirlingshire.

The county takes its name from the chief town, and the town was named from the fortress round which it grew. Dumbarton is a corruption of *Dun Breatan*, the fort of the Britons. Lennox, formerly Levenax, comes from *leamhnach*, an elm-wood. Loch Lomond, the lake of the elms, has the same origin. The word comes from the old Gaelic *leaman*, elms. Ptolemy's name for the loch, *Leamanonius Lacus*, is nearer the original Gaelic form. The river Leven, which drains Loch Lomond, takes its name from the aspirated form of the word *leamhan*, pronounced *lavan*. Sir Herbert Maxwell (our

authority for these derivations) points out that it is
interesting to find the two forms Lomond and Leven
again side by side in Fife, where the Lomond Hills
overlook the town of Leven.

2. General Characteristics.

Scotland is divided by geographers into three natural
regions: the Highlands, the Central Lowlands, and the
Southern Uplands—districts differing fundamentally in
relief, in the nature of the rocks, in scenery, in vegetation,
in industries, and in their inhabitants. These divisions
are formed by two lines that run right across the country
from north-east to south-west. One line runs from
Stonehaven to Helensburgh, and separates the Highlands
from the Lowlands; the other runs from St Abbs Head
to Girvan, and divides the Lowlands from the Southern
Uplands. These lines mark the course of two great
faults or cracks, between which the rocks of the Lowlands
have gradually sunk, leaving high hill-masses to the
north-west and the south-east. Thus when we talk of
crossing the "Highland line," it is no mere figure of
speech. By walking only a few yards we can step from
the Lowlands to the Highlands.

As the Highland Boundary Fault crosses the middle
of Dumbartonshire, it is advisable to consider briefly the
general appearance of the three main divisions of Scotland. If one looks around from the summit of a Highland
peak, one sees on all sides a chaos of mountain and

valley without any definite arrangement. As a rule, this tumbled sea of peaks rises into bare rock, sometimes rugged, splintered, and pinnacled, sometimes massive and rounded. The Southern Uplands resemble the Highlands, but in a subdued way. The hills are not so

Highland Type of Hills. Loch Long and Loch Goil

high, the scenery is not so rugged, the soil is not so sterile, the vegetation is not so scanty, and the population is not so sparse. The Lowlands are not altogether flat. In various places hill-masses rise to a height of 2000 feet above sea-level. These hills consist of great blocks of volcanic rock. They are tablelands with undulating surfaces, rising into no prominent peaks, and thus differing from both the other types. The photographs on pp. 5, 6, and 7, show very clearly the different types of scenery characteristic of the three main natural regions of Scotland.

Dumbarton is a county of contrasts. The northern half is typically Highland, the southern half is typically

Southern Upland Type of Hills

Lowland. In the north we find rugged mountains, wild and lonely glens, deep lochs with precipitous sides, and not a single town. In the south we find rich meadows, well-stocked farms, crowded towns, and river banks echoing to the clanging hammers of shipbuilders. Hardly any other county in Scotland save Lanark presents such violent contrasts. The occupations are as varied as the types of scenery. When the coalfields of Scotland began to be worked a century and a half ago, Dumbarton shared in the material prosperity that came with their development. The banks of the Clyde with their easy access to the sea, and their proximity to valuable coal and iron fields, formed an ideal home for the shipbuilding industry, and so the Dumbarton bank

Lowland Type of Hills. Flat-topped plateaus

is lined with great shipbuilding yards. The engineering industry naturally arose in the same centres, along with innumerable, subsidiary occupations. But these are not the only important vocations. The nature of the county compels diversity of occupations. In the flat lands of the south agriculture thrives, the purity of the Highland streams is the chief factor in the great dyeing industry

of the Vale of Leven, the flock-master makes what use
he can of the scanty vegetation of the north, and round
the coasts a few fishermen reap a precarious harvest
from the sea.

Long ago a Scottish king described Fife as "a beggar's
mantle fringed with gold." Nowadays this description
would apply much more accurately to Dumbarton.
Along the banks of the Clyde and the shores of the lochs
there is a narrow fringe of low ground with a dense
population engaged in many occupations. Inland the
county rises rapidly, sometimes precipitously, to high
interior hills, which are economically of little value, a
veritable "beggar's mantle." But poor only in money-
making possibilities. The breezy uplands of the
Kilpatricks or the lovely Highland hills between Loch
Long and Loch Lomond are a source of perpetual
spiritual wealth to the lover of the face of nature. Such
a one looking from the hill-tops to the border of factory
chimneys lining the north bank of the Clyde might well
reverse the epigram and exclaim, "A golden mantle with
a beggar's fringe!"

3. Size of County. Boundaries.

As far as mere size is concerned, Dumbarton is by
no means entitled to high rank among the counties of
Scotland. If we think merely of land area (neglecting
water) there are only six counties out of the thirty-three
smaller than Dumbarton Compared with Inverness,

the largest shire, our county is a pygmy, for the former
county could make sixteen of it. But if actual usefulness
to human occupations is considered, the conditions are
reversed, for Dumbarton bears a population more than
one-and-a-half times that of the broad-acred, northern
shire. Including water, the area of Dumbarton is
274 square miles, of which over twenty-eight square
miles are water or foreshore round the coast.

The county is rather odd in shape. It has a broad
base, but narrows rapidly towards the north. It is far
from easy to describe the shape with elegance and
accuracy. It may perhaps be compared to a leg of
mutton with a very attenuated shank. Neglecting for
the present the detached part of the shire, a straight
line between the north-west and the south-east extremi-
ties measures roughly thirty-four miles. From north-
east to south-west the greatest width is approximately
fourteen miles.

The boundaries are formed on the whole of strongly
marked physical features, except where in the south-east
a conflict of historical causes resulted in a political
compromise. Let us trace the boundary in some detail,
in order more fully to elucidate this statement. Beginning in the north, we find the county boundary crossing
Glen Falloch, a mile and a half north of Loch Lomond.
Perthshire lies north of Dumbarton at this point. From
Glen Falloch the boundary runs west up a little stream
called the Allt Arnan until it reaches the watershed
between Loch Lomond and Loch Fyne. Thence in a
sinuous line it follows this watershed southwards to Ben

Vane, 3004 feet above the sea.   Leaving the watershed, the boundary follows a stream for five miles to the head of Loch Long.

The next part of the boundary is simple in the extreme.   It is formed by the shores of Loch Long, the Gare Loch, and the Clyde, until Glasgow is almost

Looking up Glen Falloch from the head of Loch Lomond

reached.   At Yoker the line leaves the Clyde and runs north-east till it strikes the river Kelvin at Maryhill. It follows the Kelvin for a few miles past Garscube House and the finely wooded estate of Killermont, now a paradise for Glasgow golfers.   It may sometimes happen to an erratic golfer to drive a ball into the next county!   Leaving the main stream, the boundary runs up a tributary of the Kelvin and goes westwards, keeping a few hundred yards south of Bardowie Loch. It then sweeps round north of Milngavie, and includes also the big Glasgow reservoirs of Mugdock and Craigmaddie

The boundary next makes use of the most convenient natural feature, the Allander Water, up which it goes for some miles, and then bending somewhat to the south-west, it runs up the slopes of the Kilpatrick Hills. The line does not keep to the summit of the hills but lies near the northern border, with the result that the greater part of the Kilpatricks is in Dumbarton. Tabular, volcanic hills like the Kilpatricks, the Campsies, the Kilbarchans, the Ochils, are rarely used as county boundaries. This is probably accounted for by their plateau-like nature, there being no definite crest line to form a well-marked, political division.

The boundary runs off the northern face of the Kilpatricks by Burn Crooks. Leaving this little stream, it goes due west for a mile in order to reach another tributary of the Endrick, which it follows to its confluence with the Endrick. Then the boundary coincides with the Endrick all the way to Loch Lomond. The greater part of Loch Lomond and most of its islands form part of Dumbartonshire. From the Endrick mouth, the boundary threads its way among the islands, and then runs up the middle of the northern part of the loch. The northern tip of Loch Lomond lies entirely in Dumbarton, for two miles from the head of the lake the boundary turns eastward, reaches the shore, and climbs the mountains until it arrives at the watershed between the

Clyde and the Forth. It then keeps along the watershed for some miles, having first Stirlingshire and then Perthshire on the east. Finally it runs westwards into Glen Falloch, to the spot at which we began to trace its course.

Hitherto we have dealt only with the main part of the shire. But Dumbarton is one of those peculiar counties that are made up of separate fragments. The larger portion reaches its most easterly extension at the river Kelvin. Nearly four miles east we come again on a fragment of the shire, roughly elliptical in shape. Its long axis extends east and west for nearly twelve miles, while its maximum north and south dimension is only three miles. This part of Dumbarton includes the parishes of Kirkintilloch and Cumbernauld, formerly known as West Lenzie and East Lenzie. When one meets the phenomenon of a county split into detached portions, one may expect with confidence to find the cause to be historical rather than geographical. A geographical control makes for unity and for sub-division into natural regions. Historical causes frequently result in disintegration. Originally Lennox included not only Dumbartonshire but also much of Stirlingshire and a little of Perthshire. In the fourteenth century constant disputes occurred regarding the boundaries of the different sheriffdoms of lowland Scotland. In the reign of David II there was a rearrangement of the Dumbartonshire boundaries. The parishes of Drymen, Killearn, Balfron, Fintry, Kilsyth, Campsie, and Strathblane—in brief, the southern part of modern Stirlingshire

—were detached from Dumbartonshire, but East and West Lenzie were retained. In 1503 an Act of Parliament restored the parishes named to their original counties, which, however, was repealed six years later. Attempts were made on both sides to secure territory; and the final result of these conflicting forces was the present odd arrangement, which can be described briefly as a historical compromise.

4. Surface and General Features.

The fundamental contrast in Dumbarton is between the northern, or Highland, and the southern, or Lowland, halves of the shire. The Highland boundary runs from Kilcreggan north-east to Loch Lomond, and is a feature of profound significance. South of the Highland line the scenery on the whole is tame, the atmosphere is polluted; but, in compensation, this part of the shire supports a large and rapidly increasing population. North of the Highland line the scenery is superb, the air is clear and untainted, ugly blemishes that industrialism makes on the face of nature are absent; but, on the other hand, the land in the main is barren, and seems never likely to be capable of supporting more than the scantiest population.

In the extreme north of the county Ben Vorlich rears its lofty summit 3092 feet above the sea, while facing it across Loch Sloy, Ben Vane towers to 3004 feet. This is by far the wildest part of the shire. The rugged mountains are trenched by deep glens, which score furrows right across the watersheds in a manner most perplexing to the student of physical geography. Of these glens, the biggest is Glen Falloch, while on a slightly smaller scale Glen Loin and Glen Sloy are fine examples. The part of the county lying between Loch Long and Loch Lomond is also a region of mountain, moor, and glen; but the land is hardly so rugged and wild. The mountains of this district nowhere attain a height of 3000 feet above the sea. The most peculiar feature of this area is that the watershed is not near the middle but very close to the shores of Loch Long. Glen Douglas, Glen Luss, and Glen Fruin have their heads overlooking Loch Long, and trench a furrow right across the peninsula to Loch Lomond. The explanation of this abnormal feature will be given later in the section dealing with the rivers of Dumbarton.

The Rosneath Peninsula, between the Gare Loch and Loch Long, is Highland in character, but not nearly so rugged as the other Highland parts of the county. A band of comparatively soft slates and schists occupies most of the peninsula, and gives rise to the peculiarly smooth nature of the hill contours. The scenery of this district is particularly charming in a soft and peaceful way.

Although all of Dumbarton that lies south-east of the Highland boundary and the valley of the Leven is Lowland, the land reaches a thousand feet above sea-level. The high ground assumes the form of rolling moors covered with heather or coarse bents. The Vale of Leven is very low-lying. From Loch Lomond to the Clyde the Leven has a fall of little more than twenty feet. This low, flat land has always been subject to extensive flooding. We read that three centuries ago the very existence of the town of Dumbarton was threatened by inundations. A commission at that time reported that "na les nor the sowme of threttie thousand pundis Scottis money was abill to beir out and furneis the necessar charges and expenses in pforming these warkies, that are liable to saif the said burgh from vtter destructioune."

The most striking feature of the lowland part of the shire is formed by the Kilpatrick Hills. These hills are essentially simple in nature. Imagine a block of volcanic rock, seven miles from east to west, five miles from north to south, and a thousand feet high, and you have a conception of the main features of the Kilpatricks. The hill-sides rise rapidly from the surrounding low ground in steep, sometimes precipitous, escarpments. One of the most interesting walks in the country is that —by a delightful moss-carpeted path—along the foot of the Lang Craigs, the western escarpment of the Kil-patricks. The eye hardly knows what to choose—the ever-changing views of the Highlands and the Firth, the beauty and variety of the minerals seen at every step, and the strange forms of the crumbling volcanic cliffs,

> "The rocky summits split and rent,
> Form turret, dome, or battlement,
> Or seem fantastically set
> With cupola or minaret."

The Kilpatrick Hills do not rise into distinct summits. The top of the tableland is rough and undulating, but for miles together the variation in level is only about a hundred feet. The highest points of the plateau are just over 1300 feet above the sea. The Kilpatricks are bleak, lonely, and treeless, clothed with heath or coarse bent, with bare, rocky ribs protruding here and there. They make no arresting appeal to the eye such as is made by the imposing peaks and the serrated ridges of the Highlands. Yet they have a nameless, powerful fascination of their own. The air is clear and pure, and the miles of undulating walking afford a sense of freedom and an impression of space that true peak-climbing seldom gives.

Hills of the plateau type like the Kilpatricks are characterised by abundance of lakes. They are flat-topped with undulating surfaces, and thus the hollows of the tableland form natural receptacles for collecting the drainage of a fairly extensive area. At an altitude of over a thousand feet the rainfall is abundant, and this is also a cause of the number of small lochs on the surface of the Kilpatrick Hills. Their presence renders pleasing a type of scenery that otherwise would be bleak and monotonous.

South-east of the Kilpatrick Hills the county is low, with pleasant meadows, cornfields, woods, and occasional towns; the surface rolling and diversified, rising into ridges and dipping into valleys, but nowhere (except in the extreme east of the detached portion) rising 500 feet above the sea. The scenery of this part of the county

is naturally pleasant although tame; but many parts
are now rendered very unsightly by the traces of man's
handiwork.   A good deal of coal-mining at one time was
carried on in this part of Dumbarton, and great waste
heaps or "byngs" are found in many places.

This district is drained by the river Kelvin, which
for the most part meanders sluggishly along an almost
level bed.   All rivers have had strange and varied
histories if we could only read them aright; and the
physical history of the Kelvin is one of the most remarkable.   In the following chapter we shall give reasons for
thinking that the Kelvin once flowed in a direction
opposite to its course to-day.   In addition, there lies
far below the bed of the modern river the buried
channel of an ancient Kelvin, a far bigger river.   Borings
have demonstrated the existence of an old channel
about 300 feet below the present level.   This channel
was filled with sand and gravel in glacial times.   Enough
information, however, can be obtained regarding it to
show that the former river was deep and rapid, for the
banks are precipitous.   It follows that the whole surface
of this part of Scotland must have formerly stood more
than 300 feet higher above the sea than now.

5. Watershed, Rivers, and Lakes.

The whole of Dumbarton, with the exception of a
small part of the detached eastern portion, falls within
the Clyde drainage area.   In the extreme north-west tip
of the shire, the county boundary and the watershed

between Loch Fyne and Loch Lomond are coincident.
The watershed of the area between Loch Long and
Loch Lomond occupies a very remarkable position.
Instead of running approximately down the middle of
the peninsula, it keeps very close to Loch Long. We
shall return to this point very soon. All the drainage
of the Kilpatrick Hills ultimately flows to the Clyde,
partly direct and partly *via* Loch Lomond and the
river Leven. The low-lying part of the county north-
west of Glasgow is drained mainly by the Kelvin and
its tributaries. The detached portion of the county
lies on both sides of the main watershed of Scotland.
Most of it drains to the Kelvin and thence to the Clyde,
but the streams in the eastern part of Cumbernauld
parish ultimately join the river Forth. We always
think of Dumbarton as so essentially a west-coast
county that it seems odd to imagine any of its streams
flowing to the North Sea.

The part of the Dumbartonshire Highlands between
Loch Long and Loch Lomond is one of the "classic"
areas in the history of the study of Scottish rivers.
It was from the examination of this district that
H. M. Cadell, thirty years ago, came to the conclusion
that the Clyde now flows in a direction *opposite* to its
former course. We must consider the evidence in some
detail. A glance at a map shows us that the Gare Loch
and Loch Goil are exactly in line, and moreover that
the narrow neck of land between them is notched deeply
so as to form a comparatively low pass between the
lochs. Again, between the head of Loch Long and

Loch Lomond there is another low pass, while there is nothing in the present drainage system to account for these connecting valleys.

Now consider the peculiar position of the watershed of this part of Dumbartonshire. The Douglas Water rises on the hill-side above Loch Long, and less than

The Head of Loch Lomond

half a mile from that loch, to which it has an easy and obvious route. Yet it deliberately turns its back on Loch Long, and cuts a deep valley right through the heart of the Highlands, which are here over 2000 feet high. The Fruin Water rises less than two miles from Loch Long; yet, urged by the same mysterious influence, it prefers to trench a furrow right across the peninsula to Loch Lomond, a distance of ten miles. The map shows that the same peculiar features may be observed in the case of the Luss Water. Notice finally that the direction of nearly all the rivers of this district is towards the south-east.

All these peculiar features may be explained if we suppose that Loch Long and Loch Lomond were formed *after* the drainage system of the area had developed. The rivers then rose much farther to the west than they now do, perhaps near Loch Fyne, or, perhaps, even as far west as the coast of Scotland, for this country once extended much farther to the west. The rivers flowed south-east to the North Sea, some joining the Forth and some joining the Clyde, which then also flowed south-east. Then Loch Lomond and Loch Long were formed, and thus the head waters of the rivers were diverted west into the Atlantic Ocean. This of course explains why the Douglas, the Luss, and the Fruin rise so close to Loch Long. It also explains the low valley between Loch Goil and the Gare Loch, and the pass between Arrochar and Tarbet.

After the Clyde had been decapitated by Loch Long, part of its valley was left dry. This dry valley soon became occupied by a stream flowing *west* into the mouth of Loch Long, that is, in a direction opposite to the flow of the Clyde hitherto. This westward flowing stream pushed its head rapidly backwards until it occupied most of the valley of the old river. Thus was formed the present river Clyde. Although the foregoing account of the evolution of the river system seems highly speculative, yet so many facts point in the same direction, and the hypothesis explains so many curious features of the district, that we may be fairly confident that this account is true in the main.

In the lowland part of Dumbartonshire the Leven and the Kelvin are the most important rivers. The Leven carries off the surplus waters of Loch Lomond. The present river is too small for the big valley in which it flows, and consequently it winds sluggishly about on the floor of the Vale of Leven. The fall from Loch Lomond to the Clyde is only twenty-three feet in a distance of over seven miles. Many parts of the river are still distinctly picturesque in spite of the ugly blemishes that industrialism has made. Pennant described the river as "unspeakably beautiful," and Smollet in his *Ode to Leven Water* sings of it as

> " Pure stream in whose transparent wave
>   My youthful limbs I wont to lave "

Its purity is soiled and its transparency is dimmed. The fish that now ascend the Leven go up at the risk of their lives. It was the original purity of the stream that brought about its present state of pollution. The pure water from Loch Lomond was found to be eminently suitable for various processes in dyeing and bleaching, and this was the origin of these industries in the Vale of Leven.

The Kelvin forms one of the main boundaries of the Dumbartonshire lowlands. For several miles it bounds the detached parishes of Cumbernauld and Kirkintilloch, and also for a certain distance the south-eastern part of the main portion of the county. Although the lower part of its course is celebrated in song, yet the scenery along its valley is on the whole tame and insipid. Its tributary, the Allander, however, is a picturesque stream, and another tributary, the Luggie, of Kirkin-tilloch parish, has had its beauties sung by David Gray.

Loch Lomond is the finest lake in Dumbarton; nay more, it is the finest lake in Scotland; many would even say, in Britain. It is easily the largest lake in Scotland, being twenty-two miles long and five miles at its broadest part. The lake floor sinks far below sea-level, for while the surface is 23 feet above the sea the loch is over 600 feet deep The deepest basins are found where the loch is narrowest, and from this and other evidence most geologists believe that the lake-basin has been excavated mainly by ice. If any one county may claim the loch as its own, Dumbarton has the best claim, for most of the loch and most of the islands belong to that county.

There are about thirty islands, of which eight are of considerable size. Inchmurrin, the largest, is nearly a mile and three-quarters long. Strong castles existed formerly on several of the islands, but these are all either in ruins or have completely disappeared. The islands were the frontiers of the three chief Highland clans of this district—the Colquhouns, the Macfarlanes, and the Macgregors. Rob Roy Macgregor and the chief of the Colquhouns met on Inchlonaig to settle the monetary consideration to be paid to the famous outlaw for the Colquhoun lands to be immune from his

Loch Lomond from Luss

depredations. On Eilan Vow there was a ruined castle of the Macfarlanes where one of their chiefs lived. This solitary recluse inspired Wordsworth's poems *The Brownie's Cell* and *The Brownie*. Nowadays the loch is a favourite resort of anglers. Good loch-trout and sea-trout fishing is to be had, and even a salmon is not unknown.

The scenery of Loch Lomond is generally considered to be among the finest in Scotland. The view from the south-east looking across the islands towards the Dumbartonshire mountains, especially at sunset, is superb. The islands seem to float in a lake of molten fire, while the background of mountains, black against the glowing green or red sky, gives the desirable strong note of contrast. Dr Johnson (of course) was dissatisfied with Loch Lomond, but neither he, nor his contemporaries as a rule, could appreciate Highland scenery. Strangely enough, however, Wordsworth has not unmixed praise for the loch, and his opinions must be treated with respect even although we consider them wrong. "The proportion of diffused water" was too great, nor were the hills such as "a Cumbrian would dignify by the name of mountains." Yet Ben Lomond is higher than Skiddaw, or Helvellyn, or Great Gable, and is only eighteen feet lower than Scafell Pike, the highest peak in England. Moreover, Ben Lomond rises almost from sea-level, which is not the case with the Lake District mountains. In 1875 Queen Victoria drove along the Dumbarton shore of the lake from Tarbet to Balloch and described her experience as "perfectly beautiful."

Smollett in *Humphrey Clinker* goes so far as to say that he considered Loch Lomond finer even than Lake Geneva in Switzerland or Lago di Garda in Italy, but this is perhaps pushing local pride too far.

In olden times Loch Lomond was famed for three wonders, "waves without wind, fish without fin, and a floating island." The first is generally explained as the swell after a storm, but there is nothing wonderful in that. It may possibly have been the curious pulsation known nowadays as a "seiche," which recent research attributes to atmospheric disturbances. The loch is associated in history principally with the names of Robert Bruce and Rob Roy Macgregor. King Robert is said to have taken refuge among the islands, and to have planted yew trees on Eilan Vow and Inchlonaig in order to provide bows for his archers. Most of Rob Roy's exploits are associated with the Stirlingshire side of the lake and therefore need not be dwelt upon.

Of the other lakes of the county Loch Sloy is the most interesting. It is a long and narrow lake, occupying a deep furrow between the highest peaks in the county, Ben Vorlich and Ben Vane. The surrounding scenery is lovely and wild in the extreme. The district was the home of the Macfarlanes, and the slogan of the clan was "Loch Sloy." Several small lochs exist on the top of the Kilpatrick Hills. Their comparatively level surface combined with the heavy rainfall favours the formation of lakes. The plateau forms a fine catchment area, which can be easily utilised to supply water to the towns of Dumbartonshire; and therefore several of the lochs have been enlarged and made into reservoirs. A few years ago a new loch was created by damming Burn Crooks. This reservoir supplies Clydebank. The most important reservoir-lochs in the county are Mugdock and Craigmaddie, just north of Milngavie. In them is stored the water from Loch Katrine before

Ben Vorlich and Loch Sloy

flowing on the last stage of its long journey to Glasgow. In the detached portion of the county the principal sheet of water is Fannyside Loch, which lies on the moor of that name nearly three miles south-east of Cumbernauld. The loch is nearly a mile long but is not deep. It contains a few pike and perch but no trout.

6. Geology.

The rocks are the earliest history books that we have. To those who understand them they tell a fascinating story of the climate, the natural surroundings, and the

Types of Rocks—Sedimentary

life of a time many millions of years before the foot of man ever trod this globe. They tell of a long succession of strange forms of life, appearing, dominating the world, then vanishing for ever. Yet not without result, for each successive race was higher in the scale of life than those that went before, till man appeared and struggled into the mastery of the world.

The most important group of rocks is that known as *sedimentary*, for they were laid down as sediments under water. On the shores of the sea at the present time we find accumulations of gravel, sand, and mud. In the course of time, by pressure and other causes, these deposits will be consolidated into hard rocks, known as conglomerates, sandstones, and shales. Far out from shore there is going on a continual rain of the tiny, calcareous skeletons of minute sea-animals, which accumulate in a thick ooze on the sea-floor. In time this ooze will harden into a limestone. Thus by watching the processes at work in the world to-day, we conclude that the hard rocks that now form the solid land were once soft, unconsolidated deposits on the sea-floor. The sedimentary rocks can generally be recognised easily by their bedded appearance. They are arranged in layers or bands, sometimes in their original, horizontal position, but more often tilted to a greater or less extent by subsequent movement in the crust of the earth.

We cannot tell definitely how long it is since any special series of rocks was deposited. But we can say with certainty that one series is older or younger than another. If any group of rocks lies on top of another, then it must have been deposited later, that is, it is younger. Occasionally indeed the rocks have been tilted on end or bent to such an extent that this test fails, and then we must have recourse to another and even more important way of finding the relative age of a formation. The remains of animals and plants, known as fossils, are found entombed among the rocks,

NAMES OF SYSTEMS		SUBDIVISIONS	CHARACTERS OF ROCKS
TERTIARY	**Recent Pleistocene**	Metal Age Deposits Neolithic ,, Palaeolithic ,, Glacial ,,	Superficial Deposits
	Pliocene	Cromer Series Weybourne Crag Chillesford and Norwich Crags Red and Walton Crags Coralline Crag	Sands chiefly
	Miocene	Absent from Britain	
	Eocene	Fluviomarine Beds of Hampshire Bagshot Beds London Clay Oldhaven Beds, Woolwich and Reading Thanet Sands [Groups	Clays and Sands chiefly
SECONDARY	**Cretaceous**	Chalk Upper Greensand and Gault Lower Greensand Weald Clay Hastings Sands	Chalk at top Sandstones and Clays below
	Jurassic	Purbeck Beds Portland Beds Kimmeridge Clay Corallian Beds Oxford Clay and Kellaways Rock Cornbrash Forest Marble Great Oolite with Stonesfield Slate Inferior Oolite Lias—Upper, Middle, and Lower	Shales, Sandstones and Oolitic Limestones
	Triassic	Rhaetic Keuper Marls Keuper Sandstone Upper Bunter Sandstone Bunter Pebble Beds Lower Bunter Sandstone	Red Sandstones and Marls, Gypsum and Salt
PRIMARY	**Permian**	Magnesian Limestone and Sandstone Marl Slate Lower Permian Sandstone	Red Sandstones and Magnesian Limestone
	Carboniferous	Coal Measures Millstone Grit Mountain Limestone Basal Carboniferous Rocks	Sandstones, Shales and Coals at top Sandstones in middle Limestone and Shales below
	Devonian	Upper } Middle } Devonian and Old Red Sand- Lower } stone	Red Sandstones, Shales, Slates and Lime- stones
	Silurian	Ludlow Beds Wenlock Beds Llandovery Beds	Sandstones, Shales and Thin Limestones
	Ordovician	Caradoc Beds Llandeilo Beds Arenig Beds	Shales, Slates, Sandstones and Thin Limestones
	Cambrian	Tremadoc Slates Lingula Flags Menevian Beds Harlech Grits and Llanberis Slates	Slates and Sandstones
	Pre-Cambrian	No definite classification yet made	Sandstones, Slates and Volcanic Rocks

giving us, as it were, samples of the living organisms that flourished when the rocks were being deposited. Now it has been found that throughout the world the succession of life has been roughly the same, and students of fossils (palaeontologists) can tell, by the nature of the

Types of Rocks—Igneous

fossils obtained, what is the relative age of the rocks containing them. This is of very great practical importance, for a single fossil in an unknown country may determine, for example, that coal is likely to be found, or perhaps that it is utterly useless to dig for coal.

There is another important class of rocks known as *igneous* rocks. At the present time we hear reports at intervals of volcanoes becoming active and pouring forth floods of lava. When the lava has solidified it becomes an igneous rock, and many of the igneous rocks of this country have undoubtedly been poured out from volcanoes that were active many years ago. In addition, there are igneous rocks—like granite—that never flowed over the surface of the earth as molten streams, but solidified deep down in subterranean recesses, and only became visible when in the lapse of time the rocks above them were worn away. Igneous rocks can generally be recognised by the absence of stratification or bedding.

Sometimes the original nature of the rocks may be altered entirely by subsequent forces acting upon them. Great heat may develop new minerals, and change the appearance of the rocks, or mud-stones may be compressed into hard slates, or the rocks may be folded and twisted in the most marvellous manner, and thrust sometimes for miles over another series. Rocks that have been profoundly altered in this way are called *metamorphic* rocks, and such rocks bulk largely in the Scottish Highlands.

The whole succession of the sedimentary rocks is divided into various classes and sub-classes. Resting on the very oldest rocks there is a great group called Primary or Palaeozoic. Next comes the group called Secondary or Mesozoic, then the Tertiary or Cainozoic, and finally a comparatively insignificant group of recent or Post-Tertiary deposits. The Palaeozoic rocks are divided again into systems, and since the sedimentary rocks of

Dumbarton fall entirely under this head, we give below the names of the different systems, the youngest on top.

Types of Rocks—Metamorphic

Palaeozoic Rocks.

Permian System.
Carboniferous System.
Old Red Sandstone System.
Silurian System.
Ordovician System.
Cambrian System.

The northern half of Dumbarton is composed of ancient metamorphic rocks. No clue has hitherto been obtained to the age of these rocks, for no fossils have been found in them. All that we know with certainty is that they are older than even the Cambrian series, the most ancient of the fossiliferous rocks. The metamorphic rocks are hard and tough, and therefore they resist the denuding action of rain, rivers, frost, and ice. The northern half of the county is consequently mountainous, and forms a marked contrast in this respect to the southern half, which is hilly only where hard, igneous rocks outcrop. All the highest peaks in Dumbarton are found in the northern area of schists and gneisses. Not only is the northern half higher, but the outlines of the hills are more rugged. This is due to the same cause. The hard, folded, gnarled schists naturally give rise to a bolder and more rugged type of scenery. The characteristic scenery of the Highland schists is well illustrated by the photograph on p. 5. The appearance in hand specimens of the rocks of northern Dumbarton is shown in the photograph on p. 32.

In the extreme north of the county, the metamorphic rocks have been pierced by intrusions of igneous rock, of great scientific interest and importance. North of Loch Sloy there is a hill called Garabal Hill, composed of igneous rocks of widely different composition. They range from light-coloured granites to black, basic rocks. Yet detailed examination has shown that all the different varieties of rock have come from one molten mass by a process of separation. This area is classic ground to the petrologist.

The metamorphic rocks are not of much economic value. Among them, however, is a belt of slates that has been worked in several places. The slate-band crosses the county from Clynder to Luss, and small quarries exist at both the places mentioned. The softer nature of the slate region is reflected in the smooth contours of the hills.

The schists of the northern area are sharply separated from the younger rocks by the Highland Boundary Fault. This is one of the most important geological features in the county. It crosses the tip of the Rosneath peninsula from a point a mile east of Kilcreggan to Rosneath Bay. On the opposite side of the Gare Loch the fault is met half-way between Helensburgh and Row. Thence it runs in a wavy line to Loch Lomond, meeting the lake two miles south of Luss. The Highland Boundary Fault is a great crack along which the whole of the Lowlands of Scotland must have gradually sunk for thousands of feet. The rocks along this line have been considerably crushed, and therefore are easily attacked by the weather. The line of the fault is therefore marked in certain places by a peculiar valley, which differs from an ordinary valley in containing no river. This feature is very well seen between Rosneath and Kilcreggan, where the deep notch in the hills made by the fault is utilised by the main road. The notch is plainly visible from the opposite coast of Renfrew.

The great fault, then, separates rocks of very different kinds. South-east of it we find sedimentary rocks of the usual type. A broad band of Old Red Sandstone

rocks stretches across the county from Rosneath Point
to Loch Lomond. Except for two wedge-shaped areas
of Carboniferous rocks the Old Red extends from the
fault to the Kilpatrick Hills. This belt is very barren
of fossils.

The Kilpatrick Hills are composed of igneous rocks
that carry the mind back to one of the most wonderful
outbreaks of volcanic activity that ever took place in
this country. Hundreds of volcanoes burst into action,
hurling forth red-hot stones and ash, while molten lava
gushed from their craters, until in this way a thickness
of thousands of feet of rock was built up. From Stirling
to Campbeltown these rocks are found, forming the
Kilsyth Hills, the Campsie Fells, the Kilpatrick Hills,
the Renfrewshire Hills, and the high uplands that stretch
along the borders of Lanarkshire and Ayrshire. These
volcanic rocks are hard and resistant to the weather,
and for this reason they form hills. They were poured
out in great horizontal sheets, which explains the plateau
nature of these hills. The series was not built up in a
single outbreak. Eruption succeeded eruption, each
pouring forth a new flow of lava. Thus the rocks were
built up in a series of layers, and the resulting step-like
outline of the hills is one of the most characteristic
features of the scenery of the Clyde area. Looking
towards the south face of the Kilpatrick Hills from any
elevated point near Glasgow, one sees the escarpment of
volcanic rocks rise steeply from the softer sandstones,
not in one sweep but in a series of steps, each step marking
a different lava flow. In fact, wherever tabular hills

assume this peculiar, stepped profile, one may guess with some confidence that the hills are of volcanic origin.

In the lowlands of Scotland a steep, rocky hill frequently rises suddenly from the surrounding low ground. These hills are not very high, as a rule, but they are so steep and craggy that they form very

Dumbarton Rock

striking features of the scenery. Dumbarton Rock is a good example. It marks the place where lava and other material rose through the crust of the earth. The throat of the volcano became filled with hard rock, which was not eroded so easily as the surrounding softer sandstones; and so the "neck," or "stump," of the old volcano remains as an isolated crag. Such hills are geographically and historically of the utmost importance; for, since the earliest times, they have formed strongholds on which castles were built, and round which in time busy towns grew up. It is not a coincidence that Edinburgh Rock, Dumbarton Rock, and Stirling Rock bear the names of three of the towns most distinguished in Scottish history. It seems strange to think that these important towns owe their origin to the fact that in long past ages volcanoes happened to break out in the places where the towns now stand.

Dumbarton Rock is the best-known volcanic rock in the county, or for that matter, in the west of Scotland. A line of weakness across the neck has weathered faster than the rest, thus producing a double summit, clearly seen in the photograph on p. 36. Just over a mile to the east the rock of Dumbuck rises precipitously from the road and the railway. The columnar jointing near the top of this neck is familiar to all those who pass up and down the Clyde valley. Immediately north of Dumbuck is the rock of Dumbowie, which originated in the same way. Several other necks of a similar kind are found along the north-west borders of the Kilpatrick Hills; but being far from any main road they are not well known. In Kilmaronock parish the crag called Duncruin is very conspicuous from its isolated situation. It marks where a volcano of Carboniferous age pierced the Old Red Sandstone rocks. The last example we shall mention is interesting because of its small size, being probably the tiniest neck in the west of Scotland. It is only about twelve feet high and outcrops at the junction of Auchenreoch and Murroch glens.

Except for some small patches of Millstone Grit and Coal Measures, the rest of the county is occupied by the Carboniferous Limestone series. The massive deep-sea limestones so characteristic of this formation in England

Carboniferous Limestone rocks in Murroch Glen

are not found in the Clyde valley. The deposits were laid down in comparatively shallow water, and consist of sand-stones, shales, thin limestones, coal-seams, and ironstones. Although these rocks are not the true Coal Measures, they contain coal-seams of considerable value. Nearly all the coal and ironstone worked in Dumbartonshire come from this series. The series falls into three sub-divisions:

(a) Upper Limestone Group, containing limestones and thick sandstones; (b) Middle Group, containing several workable seams of coal and ironstone associated with sandstones and shales, but not with limestones; (c) Lower Limestone Group, containing limestones and sandstones.

After the deposition of the Carboniferous rocks, the geological history of Dumbartonshire for many ages is a blank. We have compared the record of the rocks to a book of history; and, continuing the metaphor, we may say that many of the later chapters of the work have been torn out and lost. Certainly many different systems were laid down on the Carboniferous rocks; doubtless the area was at times dry land, at times covered by the deep waters of the sea; but all the succeeding strata have been stripped away by those two all-powerful co-operators in destruction—time and the weather

The last chapter of the record tells us of the ice-age. For a long time the climate had been growing more severe. Tropical plants and animals were supplanted by temperate, and these by arctic, forms: and finally a great ice-sheet occupied all the higher parts of Scotland. Huge glaciers crept slowly down the valleys from their gathering grounds, the extensive ice-fields of the Highlands and the Southern Uplands. The glaciers have gone, but their work remains to tell their story— the grooves and scratches on the rocks, the excavation of lake-basins and the deepening of valleys, the moraines, well-nigh as perfect now as when they were thrown

Waterfall on the Dubh Uisge (Black Water)

down. One of the largest and most interesting ice-streams in the county was the glacier that formerly occupied the basin of Loch Lomond. Most geologists think that the basin of the lake was excavated by ice. Certainly along the rocky banks of the upper part of the loch the ice scoring is as fresh as if the glacier had

The Dubh Uisge
(*Ben Vane and Ben Vorlich in the distance*)

melted only a few years ago. The deepest part of the loch, too, is found where it is narrowest. At one part the lake floor sinks 600 feet below the level of the sea. On the other hand, near the foot of the loch where the ice-stream widened out and lost its erosive power the basin is much shallower. At its broadest part the loch is not much more than ten fathoms deep.

Some of the most picturesque features of the scenery are due to the over-deepening by ice of the main valley of Loch Lomond. The levels of the tributary valleys were not lowered so much, and therefore the tributary streams now fall over the precipitous sides of the main valleys as waterfalls. The well-known waterfall at

Large Erratic Block in Glen Fruin

Inversnaid originated in this way. On the Dumbarton side of the loch there is a beautiful waterfall of this kind a couple of miles above the head of the loch. This is the picturesque waterfall of the Dubh Uisge, shown on p. 40.

The glaciers from the north occasionally carried down great boulders of Highland schist and left them lying on rocks of quite a different kind. These carried boulders are known generally as "erratic blocks." Resting on the Old Red Sandstones near the mouth of Glen Fruin are two gigantic erratics of schist, which have been carried from a point many miles to the north (see the photograph on p. 42). Moraines are heaps of

Moraine near the mouth of Glen Fruin

mingled boulders, gravel, sand, and clay left by glaciers on melting. There is a long lateral moraine of special interest near the mouth of Glen Fruin. This is a ridge somewhat like a railway embankment, about twelve or fifteen feet high in places, and having the shape of a great U, the open end of which points towards Loch Lomond. The glacier which deposited this moraine was formerly·believed to have come down Glen Fruin; but, as has been recently shown, it must have been formed by a lobe of the Loch Lomond glacier which extended some way up Glen Fruin, for this event took place at the close of the ice-age when the smaller glaciers had melted. It is interesting to find this view confirmed by a water-cut gap leading from Glen Fruin to the Gare Loch. This gap occurs just at the same height above sea-level as the moraine, and was evidently the outlet to the sea of the lake formed in Glen Fruin by the natural dam of the Loch Lomond glacier.

But the Loch Lomond ice-stream interfered with the drainage of a more important river than the Fruin. It is believed to have stemmed the flow of the Clyde itself. Above Dumbarton the volcanic hills on both sides of the river approach very closely and leave only a narrow passage for the river. The glacier from Loch Lomond must have debouched on the valley of the Clyde just at this point, and thus formed a very effectual dam. The low-lying ground about Glasgow and Paisley must then have formed the floor of an extensive lake, the waters of which escaped into the Firth of Clyde by gaps in the Renfrewshire hills.

At the close of the ice-age an event occurred of fundamental importance to the future welfare of the towns on the coastal fringe of Dumbartonshire. This was a rise of the land, or (we are not sure which) a withdrawal of the waters of the sea, which converted the old sea-beach into dry land, and thus formed a narrow band of low, flat ground round the coast, an eminently suitable site for watering-places. All the coast towns of the Firth of Clyde are situated on the "raised beach," and at most of them can be seen the old sea-cliff against which the waves once dashed, now left high and dry a few score yards inland. In Dumbartonshire this is plainly visible on the roads along the Gare Loch and Loch Long. The road itself and all the houses next the sea are situated on the raised beach, which is flanked inland by low cliffs obviously of marine origin.

The volcanic rocks of the Dumbartonshire hills abound in rare minerals. Most of these are of interest chiefly to the mineralogist, but one or two are worth particular notice. The Kilpatrick Hills are famous for their zeolites, a class of minerals found among decomposed volcanic rocks. They are exceedingly beautiful, both in colour and in the shapes of their crystals. Fluor-spar, an exceedingly beautiful and somewhat uncommon mineral, is found in the county. It is familiar to many people as the "blue John" of Derbyshire, where the workers maintain that their mine is the only known locality for the mineral. Beautiful little crystals of a green or a rich purple colour have been obtained near Dumbarton. An exceedingly rare mineral called Greenockite (after Lord Greenock) has been found in the county. The crystals are yellow and lustrous, and the largest are tiny pyramids a quarter of an inch high.

The soil covering the metamorphic rocks in the northern half of the county is thin and poor. In many cases the bare rocks protrude with no covering of verdure. It is manifestly impossible to till such soil,

and agriculture is possible therefore only in the valleys.
The hill slopes are used to some extent for sheep rearing;
and much of this area constitutes fairly good grouse-
moors. In the lower parts of the shire the soil is either
a rich alluvium or a fairly stiff clay of glacial origin.
Both types can be made to bear heavy crops. Over the
volcanic rocks the soil is always very thin. In itself it
is a good soil and possesses valuable constituents; but
its thinness and the height of the land above sea-level
prevent its use in agriculture. It can be made to bear
a thick, springy turf, which when well-cut forms the
nearest approach to sea-side turf that can be found in
any inland district.

7. Natural History.

Many centuries ago the British Isles were joined to
the Continent. Where the waters of the English
Channel now ebb and flow there was dry land, offering
a free passage to the migration of plants and animals
from Central Europe to this country. Let us look for
a moment at the evidence on which this belief is based.
If the level of the sea round our shores were to sink
only 200 feet, England would again be joined to the
Continent. The English Channel and the North Sea
south of the Dogger Bank would become dry land.
The Irish Sea would become a great plain trenched
north and south by a long inlet of the sea. The chalk
cliffs of France have every appearance of having been
once continuous with the chalk cliffs of England; the

Fen district corresponds with the low-lying parts of Holland; northern Scotland is identical in geological structure with Norway. This view is supported by the facts of Natural History. The animals of England and the Continent are almost the same, although England has fewer kinds than the Continent, and Ireland still fewer. It was after the disappearance of the great ice-sheet from this country that this land-bridge existed by means of which plants and animals, including man, migrated from the Continent to Britain. But there was not time for all the animals of the Continent to migrate to this country before the formation of the English Channel and the North Sea. Thus the famous naturalist, Alfred Russel Wallace, tells us that Germany has 90 species of mammals, Great Britain has 40, and Ireland has 22. Similarly, Belgium has 22 species of amphibia and reptiles, Great Britain has 13, and Ireland has 4. The comparative poverty of animal species in Britain is most marked in the case of the mammals and the reptiles, since these do not possess the power of flight. There is not a single species of mammals, reptiles, or amphibians found in Britain that is not found on the Continent; and only one bird, the common red grouse of Scotland, does not live in continental Europe.

The mammals of Dumbartonshire resemble those of the rest of Scotland. In recent years several species have been extirpated by the farmer and the gamekeeper, while some of the smaller animals, as their persecutors lost ground, have correspondingly multiplied. Scotland is remarkably poor in bats, only four species being known as against eleven in England. Of these two species, the common bat and the long-eared bat are frequently seen in Dumbarton, but Daubenton's bat, although known in Renfrew, has not been recorded from Dumbarton. Of the Insectivora, the hedgehog, the mole, and the common shrew are widely distributed.

The wild cat is now extinct in the Clyde valley, although it lingered for a long time in the wild, mountainous country round the head of Loch Lomond. Reynard, of course, is more common than most farmers would like. He has probably thriven as his fiercer but less cunning rivals have been exterminated. The pine marten is exceedingly rare, if not extinct. The last recorded occurrences in Dumbarton were from Tarbet and Arrochar, both in 1882. The polecat is never seen nowadays, but the weasel and the stoat are not uncommon. In severe winters the latter puts on his coat of ermine fur. The badger used to be common in the Loch Lomond part of the county, but the species seems to be rapidly dying out. The otter is to be found round the coast and in the river. In recent years the common seal has been frequently seen, in the Gare Loch and other waters. The lochs of Dumbarton are occasionally visited by whales. Two bottlenose whales have been captured in the Gare Loch, and one in Loch Long. The common porpoise is, of course, a frequent visitor at all times of the year.

All the British species of rodents are to be found in Dumbartonshire except the dormouse and the little harvest mouse. The house mouse and the field mouse are everywhere common. The old black rat is now extinct. He has been pushed to the wall by his interloping relative, the brown rat. The field vole and the water vole are common, but the bank vole is little known. It has, however, been recently recorded from Luss. The squirrel was formerly unknown in the Highland part of Dumbarton, but in recent years it has increased in a surprising manner. The first species from the Loch Lomond district was killed in 1830; forty years later it was so abundant as to prove very destructive to the young plantations. Of the hare family the common species still does not belie its name, although its numbers are diminishing. The mountain hare, on the other hand, has increased and multiplied. Before 1822 no specimens were to be seen on the hills about Loch Lomond, where they are now abundant. The rabbit shares with the brown rat the distinction of being the commonest of all the mammals.

It would take far too long even to mention the names of all the birds found in Dumbarton, and therefore we shall confine ourselves to a few comments on the most uncommon species. It should be noted that the Loch Lomond district is one of the best places in the west of Scotland in which to study bird life. The siskin is very rare in the Clyde valley. One of the few trustworthy records of this species is from Luss. Another extremely uncommon bird is the tree-sparrow, which, however, has been reported from Arrochar. The rare parrot-crossbill probably nests in the Loch Lomond district. The jay is now nearly extinct in the Clyde valley. A pair or two survive in Ayr, but in Lanark and Renfrew they have disappeared. Yet they are still to be met with occasionally in the Loch Lomond part of Dumbarton. The distribution of the magpie in the west of Scotland is rather odd. It is not uncommon in Ayr, whereas in Dumbarton it is rarely seen. On the other hand, while the hooded crow is rare in Renfrew and Lanark, it is quite well-known in the Loch Lomond district. The night-jar is often heard on the Dumbarton moors near Loch Lomond, but it is seldom that one sees the blue flash of the kingfisher as he darts along a stream. The rare hoopoe has been recorded from the Clyde valley only about half a dozen times, one of the occurrences being at Cardross.

There are several quite common species of owls found in Dumbarton. The snowy owl, however, is a very rare, winter visitor to the west of Scotland. It has only twice been recorded from Dumbarton. Of the falcon family the golden eagle is the most interesting. It probably still builds its eyrie on the mountainous borders of Dumbarton and Argyll. The kite was common in Dumbarton in the first half of the nineteenth century, but is now unknown. The same may be said of the osprey. It used to nest in the Loch Lomond district, but seems now to be extinct in the Clyde valley. There are only two reliable records of the marsh harrier in the Clyde area, both from Dumbartonshire.

Many members of the goose family are to be seen in the county, particularly about Loch Lomond. Several species of swans are winter visitors to this district. A few species of ducks, some very rare, are also found in the same region. The smew and the gadwall are excessively rare. The former has been recorded only from Loch Lomond; of the two authenticated records of the latter in the Clyde valley, one is from Dumbarton. The wood pigeon is common, and the much rarer turtle dove is known. The grouse, the pheasant, and the partridge are common, and the ptarmigan is occasionally seen. The oyster catcher breeds at Loch Lomond. The snipe is common in marshy places, and the mournful cries of the lapwing and the curlew are everywhere heard on the moors. There are several varieties of sandpiper, including the very rare wood sandpiper and green sandpiper. On the islands of Loch Lomond many species of terns are found, including some rare kinds. The arctic tern is said to have nested on Inch Moan. The Loch Lomond islands, too, are inhabited by colonies of various species of gulls.

Most of the British reptiles and amphibians are found in Dumbartonshire. The lizard may often be seen on a hot day frequenting dry, sunny places, such as stone-heaps, walls, or ruined buildings. The blind-worm or slowworm is to be found among dead wood, decayed leaves, or stone-heaps, generally preferring a dry situation. It is not a true snake, although it is often mistaken for one. It is in reality an inoffensive, timid, and perfectly harmless creature. Of the true snakes the adder or viper is our only representative. It is the only poisonous reptile in the country. To the healthy adult its bite is seldom fatal, although death has resulted in the case of children and infirm persons. The adder loves dry, warm places among ruins, under fallen trees, or on sunny banks. The frog, the toad, and the newt are everywhere abundant. The scarce natterjack toad is not known, but the very rare crested newt has been seen in a quarry hole at Helensburgh

The plants of Dumbartonshire are fairly representative of the whole of Scotland. The summits of Ben Vorlich and the neighbouring mountains exhibit examples of most of the peculiar Alpine plants found in Scotland, except a few of the purely arctic forms that are confined to the most lofty summits such as Ben Nevis or Ben Lawers. The middle slopes of the Dumbarton mountains and the Highlands between Loch Long and Loch Lomond afford typical examples of the flora of the Scottish grouse-moors. The old Caledonian forest probably existed over many areas that are now bare of trees. The existing woods of Dumbarton have practically all been planted by man. Deciduous trees are best developed in the lowland parts of the county, and many fine examples may be seen in the estates of the larger mansion-houses, notably Balloch Castle, Killermont House, Woodbank, and Tullichewan Castle. Dumbarton possesses more than its share of very tall trees. A tree over a hundred feet high is a veritable giant. In the list of the tall trees of the West of Scotland compiled for the British Association meeting of 1901, there are only ten trees recorded over a hundred feet; of these four are in Dumbarton. At Catter House there is a huge Scotch elm 101 feet high and over 13 feet in girth. A gigantic beech at Craigton attains a height of 104 feet and a girth of 14½ feet. A black poplar at Catter House is 102 feet high, while the tallest recorded tree is a poplar at Luss, which reaches the great height of 105 feet. At Rosneath two gigantic silver firs are

The Yew Tree Avenue, Rosneath

worthy of special mention, and the avenue of yew trees is the finest in the country.

Limitations of space prevent us going into detail regarding the plant life of the county, but a few words may be said about the mosses and lichens, for Ben Vorlich is one of the most famous localities in Scotland for these plants. This mountain forms the centre of an area where the genus *Campylopus* is particularly well

developed. These mosses give a characteristic appearance to the whole vegetation of this area. Forty years
ago a moss (*Didymodon recurvifolius*) was discovered on
the slopes of Ben Vorlich. At that time this was the
only known station in existence. On this mountain,
too, and about the same time, a lichen was discovered
which was new to British science. On Dumbuck Hill
there is found a crowd of mosses of the genus *Grimmia*,
some of which are exceedingly rare and beautiful. It is
remarkable that on a number of detached hills of the
same volcanic nature that stretch across Scotland from
Dumbarton Rock to Stirling Rock and thence to
Arthur's Seat, the same group of mosses is found, and
found nowhere else.

The Kilpatrick Hills are mainly moor or marsh.
In autumn they are purple with the flowers of the ling
and the heath. The milk-wort, the bog asphodel, and,
in wetter parts, the cotton grass, are abundant. In the
marshes, also, the butter-wort and the sundew set their
traps for unwary insects. All summer the grassy
slopes of these hills are bright with the tiny, yellow
flowers of the tormentil, and the gaily coloured mountain
pansy The sheltered hedge-rows of the Gare Loch and
Loch Long are rich in flowering plants, but the south-
eastern parts of the county are generally covered with
boulder-clay, which gives a stiff, cold soil that is not
favourable to variety of plant life.

8. Around the Coast.

It is not easy to say exactly where the river ends and the estuary begins; but as there is a noticeable widening of the channel below Bowling, we may make that town the starting-point of a peregrination round the Dumbartonshire coast. The scenery in the neighbourhood of Bowling is particularly pleasing, and has given inspiration to several of Scotland's most famous artists   Sam Bough and Horatio McCulloch did some of their finest work here. Just below Bowling, on a little rocky promontory, stands all that is left of old Dunglass Castle, a former stronghold of the Colquhouns. On the highest point of the promontory a conspicuous obelisk commemorates Henry Bell, who designed the famous *Comet*, the world's first passenger steamer. The Kilpatricks come very close to the river at Dunglass, and a mile lower down one of their outposts—Dumbuck Hill—approaches the Clyde so closely as barely to leave space for two railway lines and two main roads. Dumbuck Hill rises sheer from the road. It is finely wooded, and is one of the most picturesque of the smaller hills of the county. The view from the summit is superb. The hill is formed of igneous rock which on cooling has cracked into columns similar to those so well known in Staffa and at the Giant's Causeway, although they are of course not nearly so perfectly formed. Dumbuck Hill completely com mands the town of Dumbarton, and was occupied in 1745 by the troops of Prince Charlie.

Outward Bound

(Dumbartonshire Hills in the background)

A mile beyond Dumbuck we reach Dumbarton, not an inspiring town either in itself or its immediate surroundings. As we cross the Leven and leave the drab town behind us we find both road and railway running close to the shore. They make use of the levelled surface of the old raised beach, while inland we may see low cliffs that marked the shore-line of a former age. When the tide is in, the view seaward is charming; but it is spoiled at low tide by a great expanse of mud flats. The navigable channel keeps close to the opposite shore. Straight across the estuary, here two miles wide, we see the smoke of Port Glasgow, and perhaps hear the faint clatter of hammers beating on the steel skeletons of ships Just above us on the slopes of Cardross Hill stood Cardross Castle, where Robert the Bruce spent his last days. A mile farther on we pass through the modern village of Cardross, a pretty, straggling place. Before reaching Craigendoran we see on our left a curious, wooded headland joined to the mainland by a narrow neck. This is an example of a "tied island." Ardmore was formerly an island but the tidal scour has swept sediment between the island and the mainland, until the island was "tied" to the land, and thus converted into a peninsula. Craigendoran, one of the three main starting-places for the steamer-traffic of the Firth, is now merely the east-end of Helensburgh. The latter town is purely residential. The arts have always flourished in Helensburgh. Music is cultivated to some purpose, and a town favoured by artists of the calibre of Colin Hunter, Milne Donald, Sir Alfred East, and Sir James

Guthrie may claim to have done something to foster painting. As we round Cairndhu Point we come fairly into the Gare Loch. Between Cairndhu and Castle Point opposite, the loch contracts; between Row Pier and Rosneath Pier there is another very marked narrowing; but between these two constrictions the loch

Training Ship *Empress*

widens into a sheltered basin that is a haven of rest for many yachts. Pert little racers, big cruising schooners, and stately steam yachts swing at their moorings—a goodly sight for all who love the sea; but the eye leaves them and dwells on the *Empress*, now a training ship, formerly H.M.S. *Revenge*, one of the few remaining representatives of the old "wooden walls." Row (pronounced Roo) takes its name from *rudha*—a long, low point.   The suitability of the name is obvious when one sees the long spit of shingle that juts far into the loch, leaving a very narrow passage between the end of the point and the opposite shore, which is also salient. Through this passage at some states of the tide, the water rushes like a mill-race, and the strait is disturbed by confused, short, choppy waves, which make navigation by a small boat difficult and dangerous.   The strait is known locally by the expressive name of "The Jumble."

Rounding the point at Row we enter the straight, upper portion of the Gare Loch, a stretch about five miles long.   Almost land-locked, and well-sheltered by hills, the Gare Loch is the safest boating loch on the Firth of Clyde.   It is a haven of rest also for large craft, which lie anchored waiting for better times.   For the Gare Loch is a barometer of the shipping trade.   When freights are low and scarcely worth taking, there may be twenty or thirty big steamers in the loch, swinging idly at their cables, rusting their hearts out for a sight of foreign ports.   But when trade improves, anchors are raised and the ships steam gaily away, leaving the loch empty, except, perhaps, for some old worn-out derelict, fit only for the "breakers."   The sides of the Gare Loch are lined with fine mansions, for the climate is unusually mild, and the locality is not too far away from the large towns of the Clyde valley for daily travel.   Indeed, there is an air of quiet, unostentatious wealth about the Gare Loch that no other Clyde loch can boast.

The only opening that indents the straight sides of the Gare Loch is Faslane Bay, near which was Faslane

The Gare Loch in pre-war times

Castle, an ancient seat of the Earl of Lennox, now represented only by a grassy mound. Not far away are the ruins of a pre-Reformation chapel. Twenty years ago the head of the loch was frequently visited by ocean liners for the purpose of adjusting their compasses. Five huge, red buoys anchored near the head of the loch were used in the operations, which involved much warping, turning, signalling, and shouting, and were a source of never-failing interest to the younger inhabitants of Garelochside. The west side of the loch need not detain us long. It has undoubted beauty of a quiet kind. One or two sleepy, little villages with quaint names— Mambeg, Rahane, Clynder—front the shore. Rosneath, however, is the most interesting place on this side. Its etymology is frequently obscured by the erroneous spelling *Roseneath*. The first part is certainly *ros*, a headland, and the second part probably commemorates Nevydd (pronounced Nevith), an ancient martyred bishop. The neighbourhood boasts of a magnificent avenue of yews, perhaps the finest in Britain. Looking into Rosneath Bay is one of the principal seats of the Duke of Argyll, for we are now in undisputed Campbell country. The present palace was built in 1805. It is an imposing building in the Italian style with Greek modifications. The picturesque inn at Rosneath pier was designed by the Princess Louise. Sir Walter Scott lays the scene of the closing chapters of *The Heart of Midlothian* at Rosneath, but he makes one little slip in topography, for he calls it the "picturesque island of Rosneath."

The main road between Rosneath and Kilcreggan
does not go round the southern tip of the peninsula,
but takes a short cut across, making use of a deep, dry
valley that marks the line of the great Highland Boundary
Fault.   North of the valley we find hard grey schists;
south of it we see the bright red rocks of the Old Red

Rosneath Castle

Sandstone.   The valley has been formed by the rapid
weathering of the broken and crushed rock along the
fault-line.   As we descend to the village of Kilcreggan,
we see on our left some innocent-looking, green mounds.
These mask the deadly batteries of Portkil Fort.
Should a hostile man-of-war by some chance escape
the guns of Ardhallow Fort lower down the Firth, it would
get a warm reception off Kilcreggan from Fort Matilda

on one side and Portkil Fort on the other. The view from Kilcreggan is particularly fine One looks right down the Firth. In summer the sparkling, blue waters dotted over with the white sails of pleasure yachts, and the passenger-steamers with their distinctive colour-notes in their funnels give animation to the scene. The white buildings of the Cloch Lighthouse are prominent, and even Toward Light can be seen in the distance. Still farther off is the noblest background in the Clyde valley, the splintered, granite peaks of Arran.

On leaving Kilcreggan we round Barons Point, and are now at the mouth of Loch Long. The road keeps along the very edge of the sea, and one beautiful view succeeds another Cove is a quiet little place with a number of fine mansion-houses. Its steamer pier is the last we shall see until we reach Arrochar at the very head of the loch. The road itself stops short at the small pier of Coulport, three miles past Cove. The loch now becomes narrower and almost perfectly straight. Its floor is hollowed out in deep basins. Its rocky sides plunge steeply, often precipitously, into the water, making a road impossible except by cutting it from the hill-side. If we wish to continue, we must take to the heather. This type of coast-line—interrupted frequently by delightful, little sandy coves—continues for several miles until we reach the little fishing village of Portin-caple. For many years Charles Bradlaugh spent his summer holidays in this charming, little place. Just opposite Portincaple Loch Long sends off a branch (Loch Goil) to the north-west. The peninsula between the two lochs is exceptionally rugged, and is therefore ironically known as "Argyll's Bowling Green." (See p. 5.) The southern part of the peninsula—the estate of Ardgoil—belongs to the Corporation of Glasgow, having been presented to the city for use as a public park.

Loch Long from The Cobbler
(*Showing the Croe Delta*)

A mile past Portincaple the main road again reaches the shore. The West Highland Railway line runs parallel to it, but up the hill-side a little. On some parts of the line it is as well for timid passengers not to look out of the windows, as there seems to be a sheer drop from the carriage right down into Loch Long. The upper part of the loch, being straight, deep, and little frequented, is now used by the Admiralty as a practice-station for testing torpedoes. A mile before we reach Arrochar, we see opposite us one of the prettiest examples of a delta to be seen in Scotland. It has been formed at the mouth of Glen Croe. Behind it rises the fantastic, gashed ridge of The Cobbler. Arrochar is a quiet little

Looking across Loch Lomond to Tarbet and
The Cobbler (right background)

village at the head of Loch Long. It makes a splendid centre for the tourist or the mountaineer. Between Arrochar and Tarbet on Loch Lomond there is a deep pass, eroded perhaps by a river whose head-waters were tapped and diverted by Loch Long. The origin of the name Tarbet is significant. It comes from the Gaelic *tarruin bad*, meaning draw boat. When Haco (or Haakon) of Norway invaded Scotland in the thirteenth century he sent his son-in-law Magnus, King of Man, to harry the Clyde lochs. Magnus sailed to the head of Loch Long, and then drew his boats over the low isthmus to Loch Lomond. He sailed down this loch, burning, plundering, and slaying, and then returned by the way he had come.

9. Weather and Climate.

The weather of Britain depends largely on the distribution of atmospheric pressure over our islands and over the seas and lands surrounding them. When the barometer is high, we expect good weather; and when the barometer is low, we expect wet and stormy weather. These two types of weather correspond respectively to a condition of high atmospheric pressure or anticyclone, and a condition of low pressure or cyclone. Look at the figure on p. 67, which shows the pressure and direction of the wind on a day in winter. The dotted lines are drawn through places that have the same pressure, and are called isobars. The figure represents a type of weather conditions very frequently experienced in this country. The winds are strong, and swirl round the centre of lowest pressure in great spirals with a direction opposite to that of the hands of a clock. Rain is frequent and abundant, and in winter there is a rise in temperature. Look now at the figure on p. 68, which shows an anticyclone over the British Isles. In this case the distance between the isobars is much

greater than in the former figure; in scientific language
the barometric gradient is much less. The winds,

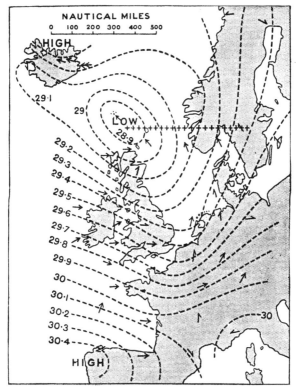

Pressure Chart illustrating a Cyclone. November 22, 1908

therefore, are very light, and this is characteristic of
anticyclones. There is a gentle circulation in the

same direction as the hands of a clock. Fine, clear,
sunny weather is usually associated with an anticyclone

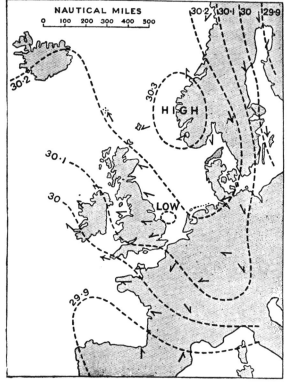

Pressure Chart illustrating an Anticyclone.   November 6, 1908

In summer the weather is sunny and hot; in winter it
is clear, cold, and frosty.   In low-lying ground and in
towns there is a tendency to mist and fog.

Speaking generally, we may say that the weather of this country is controlled by three fairly permanent pressure centres. There is a low-pressure area south of Iceland, an Atlantic high-pressure area about the Azores, and a Continental area in eastern Europe and western Asia that is high in winter time, and low in summer time. The frequent cyclones that invade our coasts seem to be smaller eddies thrown off by the permanent Icelandic low-pressure area. In winter, as a rule, the Icelandic and the Continental centres predominate. They are then working in harmony, as the British Isles are between them. The tendency of both centres is to draw the air in a great swirl between them from south-west to north-east. Therefore we find that in winter south-west winds predominate in Scotland. In summer the Atlantic high-pressure centre has more influence. This area with its accompanying fine weather is now at its most northerly limit, for to some extent it moves north and south with the sun. In addition, the Icelandic and the Continental centres are no longer in harmony, for both are now cyclonic, and Britain lies between them. The anticyclone occasionally spreads over these islands, reaching the south of England frequently, but not so often extending to Scotland. The Atlantic anticyclone tends to draw the winds more to the west, sometimes even to the north-west.

We have stated that the winds shift from south-west to west according to the season, and we have offered an explanation of this shift. Let us see now if actual observations, over several years, confirm the statements we have made. We shall show the average wind directions for sixteen years, from 1893 to 1908 inclusive. The wind conditions are practically the same all over the Clyde valley, and we shall therefore make use of the records of the University Observatory, Glasgow, for there are no other data in the west of Scotland comparable in fulness and accuracy. Instead of giving numerical tables, we have expressed the results as diagrams from which the prevailing winds may be seen at a glance. Along each of the eight principal points of the compass we mark a distance proportional to the percentage of days on which the wind blew from that direction, and so get a wind-rose or wind-star, the longest arms of which show the winds that blow most frequently. The top figure shows that the winds of winter are chiefly from the south-west, and the second figure shows that the winds of summer are chiefly from the west. In late spring and early summer, easterly winds are fairly common. This results from the fact that in spring, cyclones frequently travel along the English Channel or the Bay of Biscay. A little consideration will show that this must result in cold east winds from the Russian plains being drawn over the British Isles. The third figure on p. 71 shows the prevailing winds for the whole year. West and south-west winds are clearly the most common. In many parts of the country the trees are silent but reliable witnesses to the same fact. They grow with their branches pointing east or north-east, away from the wind. The branches of the tree shown in the photograph

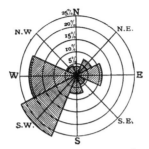

Wind Rose showing the frequency of the winds from the
eight principal points of the compass during January

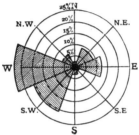

Wind Rose showing the frequency of the winds from the
eight principal points of the compass during July

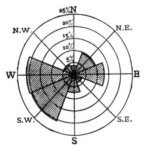

Wind Rose showing the frequency of the winds from the
eight principal points of the compass throughout the year

on p. 72 (taken on a still day) point exactly north-east.

It is still quite generally believed that storms are more frequent and violent at the time of the equinoxes than at other times. The phrase "equinoctial gales"

Tree deformed by prevailing south-west wind

is heard so frequently that people come in time to believe that September and March are especially stormy months. There is no foundation for the belief. The equinoctial gales are mythical. In order that there can be no doubt about the matter, let us take a period of forty years, 1868–1907, and note the number of gales in each month, recording all storms over forty miles an hour. These are the numbers for each month during the whole of that time:

Jan.	Feb.	Mar.	Ap.	May	Je.	Jy.	Au.	Sep.	Oct.	Nov.	Dec.
50	42	36	11	5	2	2	5	10	15	27	39

These figures show clearly that storms are most frequent in winter and least common in summer. The maximum number occurs in January, and the number steadily decreases till June and July, then steadily rises to January.

The prevailing south-west winds of this country have much to do with the mild climate that we enjoy. These winds blow the warm surface waters of the Atlantic Ocean around our shores, and we are therefore favoured with winters milder than those of any other country in the same latitude. Thus while the *mean* winter temperature of every part of the British Isles is a little above the freezing-point, the mean winter temperature of parts of Canada in the same latitude is nearly forty degrees below freezing-point. We must thank the south-west wind for the fact that outdoor work can be carried on with so little interruption in Britain, and that our harbours are navigable throughout the winter. It was formerly believed that our good fortune as regards climate was due to the beneficent influence of the Gulf Stream, but this explanation has been abandoned in recent years. The warm currents washing our coasts in winter are surface drifts caused by wind.

The temperature conditions of Dumbartonshire are

similar to those of the other countries of the western
sea-board of Scotland. The summers are cooler and
the winters milder than on the east coast. The mean
temperature of Cardross, for example, is 38° F. in
January and 58° F. in July. This gives a *mean* annual
range of temperature of 20° F. The mean annual

Curling on Loch Lomond during the great frost
of February 1898

range for Edinburgh is 21° F. and for London is 26° F.
These ranges are of course small compared with the
differences of temperature observed in the interior of
continents. In the heart of Asia, for example, a mean
annual range of 100° F. is not unknown. The chief
differences of temperature between different parts of
Dumbartonshire are caused by differences of altitude.

It is an observed fact that as we rise above sea-level the temperature falls approximately 1° F. for every 300 feet we ascend. Now between the highest and the lowest parts of Dumbartonshire there is a difference in altitude of 3000 feet, therefore some parts of the county are roughly 10° F. colder throughout the year than others.

The records of rainfall for the county are neither so complete nor so trustworthy as one could wish. All the older statistics regarding weather phenomena must be accepted with grave hesitation. In many instances they can be shown to be utterly unreliable. A good example of the reckless statements of the older writers is given in the account of Dumbartonshire in the *New Statistical Account* of 1845. The writer attributes the rainfall of the county to the "broad stripe of coal and lime, and the great masses of iron in beds or balls." He goes on to say that the rainfall of the opposite coast from Greenock to Ayr is much greater than any corresponding district in Scotland, a statement which of course is most inaccurate. He attributes the rain to "those subtile and generally invisible agents, the magnetic or electric fluids," which are attracted by the iron in the ground.

The following table is compiled from the annual volumes of *British Rainfall*, which gives the most trustworthy data concerning the rainfall of Dumbartonshire that we can obtain. The stations have been selected that they may, as far as possible, be representative of the whole county. The figures given show the average rainfall for the ten years 1903–1912 inclusive.

Station	Height above sea-level	Rainfall
Dumbarton	9 feet	48·3 inches
Helensburgh	18 ,,	54 ,,
Rosneath	25 ,,	63·2 ,,
Garelochhead	30 ,,	67·4 ,,
Arrochar	50 ,,	89·1 ,,
Kilpatrick Hills	400 ,,	48·7 ,,
Kilpatrick Hills	912 ,,	59·9 ,,

The first five places on the list are all on the sea-coast, and the stations are all near sea-level, yet the rainfall of Arrochar is nearly double that of Dumbarton. The explanation obviously is that whereas Dumbarton is situated in a low-lying district, Arrochar is in the Highlands and is surrounded by mountains. We invariably find that rainfall increases with height above sea-level. The mountainous parts of Britain are the rainiest, the flattest parts are the driest. The last two stations in the table illustrate the same point. They are both stations in the Kilpatrick Hills and can therefore be directly compared. The more elevated station has much the greater rainfall. The explanation was partly contained in a former paragraph where it was stated that for every 300 feet of increase in altitude the temperature falls 1° F. When moisture-bearing winds reach elevated land they are forced to rise and are consequently cooled. But cold air can hold less moisture than warm air, and therefore the moisture is frequently

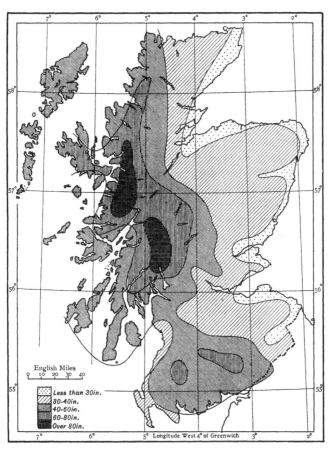

Rainfall Map of Scotland. (After Dr H. R. Mill)

dropped as rain. We may sum up the case for Dumbartonshire by saying that the Lowland parts of the
county have a moderate rainfall, while the rainfall of
the Highland part is high.

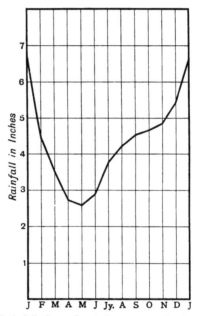

Rainfall throughout the year at Cardross

Let us consider next the distribution of rainfall
throughout the year. The figure on p. 78 is a graph
showing the variation in the rainfall from month to
month at Cardross. It represents the average readings
of twenty-five years and is similar in character to the

curves obtained for other stations, and therefore we may take it as typical of the county as a whole. It is evident that the winter months are by far the most rainy. The wettest month is January, the driest months are April and May. This is generally the case in the west of Scotland. It is connected with the fact stated earlier in the chapter that cyclonic disturbances then tend to pass south of England. The marked rise in the rainfall in July and August is only too familiar to holiday-makers in the west of Scotland. As regards length of daylight, dryness, and hours of bright sunshine, June is undoubtedly our ideal month of summer.

10. The People — Race, Language, Population.

The earliest inhabitants of Britain of whom we have any definite knowledge used roughly chipped stone tools and weapons, and are therefore called palaeolithic (ancient stone) man. They were contemporaneous with the wild horse, the mammoth, and the reindeer, for they have left drawings scratched on bone or slate, of all these animals, and very well executed drawings they are. This race probably reached England from the Continent by land before the formation of the English Channel. Although relics of palaeolithic man are abundant in England, most authorities agree that there is no evidence for their presence in Scotland. They were succeeded by a totally different people which used finely chipped and polished tools of stone, whence the

race is called neolithic (new stone) man. Relics of
neolithic man are found all over Scotland, and form the
first traces of human occupation in north Britain.
The neolithic race had long skulls, dark hair and com-
plexions, good features, and were short in stature, say
about five feet four inches in average height. They are
known sometimes as Iberians, and have few affinities,
either in appearance, culture, or language, with the
Celts and Teutons.

Later on Scotland was invaded by Celtic tribes,
who were tall and had broad skulls. When historic
records commence, nearly all the people of Scotland
spoke some dialect of the Celtic language, with, however,
a number of non-Aryan words and peculiarities of syntax.
From this test of language it has been concluded that
the original Iberian inhabitants were driven out or
exterminated by the invading Celts. This view, however,
is contradicted by anthropology. At the present time,
most of the people of Scotland are long-headed, and therefore cannot possess much Celtic blood, for the Celts
were broad-skulled, and the shape of the skull is one of
the most constant and valuable physical characteristics
of a race. Again, the Scottish Highlanders and other
Celtic-speaking races of Britain are very much darker in
complexion than the people of other districts. One is
forced to the conclusion, therefore, that the aboriginal
Iberian stock was not eliminated by the Celtic invaders.
It seems, indeed, to have been the reverse. The Iberians
absorbed the Celts without serious dilution of their
original characteristics. On this view the Celts were

merely a predominating and ruling caste, who imposed their language and culture on the conquered tribes, but had little effect on their racial nature. No definite agreement on these points, however, has yet been reached.

At the beginning of the Christian era the inhabitants of our county were Brythonic Celts, and the Rock of Dumbarton was the northmost stronghold of the British Kingdom of Strathclyde. From the fifth century Strathclyde had as its neighbours on the north-west Goidelic Scots, who had come from Ireland and occupied Argyll, and on the east Angles from across the North Sea. These various people met in peace and in war, and intermingled. The inhabitants of Dumbartonshire, then, are a curious mixture. In the Highland parts the blood of the ancient Iberian or pre-Celts still predominates, in the Lowland parts the Teutonic stock is the prevailing one, but there is some admixture of the two, and both races are also crossed with the tall, fair, broad-skulled Celts.

With very few exceptions the place-names of Dumbarton are either Anglo-Saxon or Celtic. Naturally in the Highland part of the county the names are almost entirely Gaelic. These names often describe in a vivid way the most outstanding characteristic of the hill or river or cape. Thus a jutting-out hill near Garelochhead is called *The Strone*, "the nose," the village at the mouth of the Gare Loch is Row, from *rudha*, "a low point," *Ben Bhreac* is "the dappled hill," *Ballevoulin* is "the mill-homestead," *Knockvadie* is "the hill of the wolf," *Craigendoran* is "the rock of the otter." In the lowland part of the county the place names are partly of Celtic and partly of Anglo-Saxon origin. The natural features (hills, rivers, etc.) are generally Celtic, while town or farm names are often English, e.g. *Bearsden*, "the wild-boar's lair," *Summerston*, *Dougalston*. Of course there are numerous exceptions to this statement. *Dumbarton* and *Kirkintilloch* are pure Gaelic, and *Thief's Hill* is plain English. Some of the names, such as *Renton*, *Alexandria*, *Helensburgh*, are of quite recent origin, and owe their origin to a desire to perpetuate the memory of certain people. Names of Danish origin are rare in the county.

In size, only six Scottish counties are smaller than Dumbarton. In population, however, the case is different, for Dumbarton ranks ninth among the thirty-three counties. The total population in 1911 was 139,831—69,718 males, 70,113 females. In recent years the growth of population in Dumbarton has been exceptionally rapid. The county now contains over six times the number of inhabitants it had at the beginning of the nineteenth century, and during the ten years 1901 to 1911 the population increased more rapidly (in proportion to numbers) than that of any other shire in Scotland. The explanation of course is to be found in the marvellously rapid development of the part of the county bordering the Clyde between Yoker and the town of Dumbarton. If we compare Dumbarton with the neighbouring county of Argyll, the contrast is very striking. In 1801 the population of Argyll was four times that of

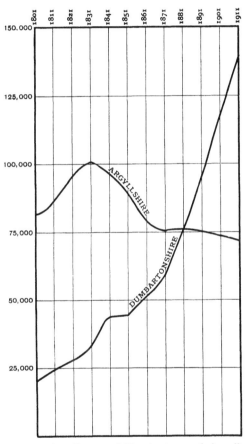

Population Curves of Dumbarton and Argyll

Dumbarton. At the present time the population of the latter county is twice that of the former. Dumbarton too, can boast of containing the town that has grown more rapidly in the past ten years than any other town in the British Isles. Between 1901 and 1911 Clydebank showed the phenomenal increase of 79·7 per cent. in its population. Of the parishes Old Kilpatrick is the most densely populated, having 360 persons per 100 acres, while Arrochar and Luss are most sparsely populated, having only two persons per 100 acres.

The foreign element is not particularly strong in Dumbartonshire, not nearly so much as in Lanarkshire, which contains more than half of all the foreigners in Scotland. In all there are 324 aliens in the county. The sweet tooth of youth probably accounts for the fact that there are more than twice the number of Italians than of any other nationality. Germans and Spaniards rank next in numerical order. In 1911 there was one Asiatic and no Africans. In the Highland part of the county the Gaelic language still keeps its hold fairly well. Roughly speaking, between five and six thousand people live north of the Highland line in Dumbarton and of these over three thousand are able to speak Gaelic. They are practically all bilingual. Only one Gaelic-speaking person is unable to speak English.

The industries employing the largest number of men are shipbuilding and engineering. Nearly 27 per cent. of all the occupied men of the county are engaged in shipbuilding, and 18·5 per cent. in iron manufactures. Coal-mining ranks next in the number of men employed, and then the bleaching and dyeing industry. The former claims 5·8 per cent. of the workers and the latter 5·4 per cent. The agricultural element is not very strong, the percentage being 4·2. Of the women 39,000 are supposed to have no occupation, compared with 15,000 who work. This is explained by the fact that household duties, no matter how heavy, receive no salary, and are therefore not considered "work" by the census. Of the special occupations followed by women domestic service is most popular, and the bleaching and dyeing industry ranks second. Practically half the people live in two-roomed houses, 20 per cent. live in three-roomed houses, and 8·5 per cent. live in one-roomed houses. The rate of mental infirmity is, according to the census of 1911, significantly less in Dumbarton than in Scotland as a whole.

11. Agriculture.

Little real progress in Scottish agriculture was made until the eighteenth century. The custom of short leases was partly to blame. One of the greatest hindrances to progress was the deplorable system of "runrig," which was almost universal. In this system each field was divided into ridges or rigs usually about twenty feet wide. Each ridge was farmed by a different tenant. Only the crown of the rig was ploughed, leaving the borders undrained and overgrown with weeds and nettles. As mutual help was needed for many farming operations involving the use of animals and implements, the system depended for smooth working on harmony and co-operation, which were generally lacking. Jealousy and bickering were even worse enemies to the crops than the inclement Scottish weather. The runrig system practically became extinct nearly a century ago, but even now it is not unknown in farms in Shetland.

At the beginning of the eighteenth century the land was undrained, the fields were not enclosed by fences, and means of communication were primitive in the extreme. The horses and oxen were kept on such short commons through the winter that when yoked to the plough in spring they fell into bogs and furrows through sheer weakness, and this in spite of the fact that in order to render them more fit for work they had first been copiously bled! The harrows, made entirely of wood, "more fit," as Lord Kames said, "to raise laughter than to raise soil," had been in some districts dragged by the tails of the horses, until the barbarous practice was condemned by the Privy Council. Gradually the old methods were supplanted and new crops were introduced. The land was let in large holdings, the material taken from the soil was replaced by plentiful manuring, and by these means along with judicious crop rotation exhaustion of the soil was avoided, systematic stock-rearing was practised, turnips and potatoes were largely introduced, until by the middle of the nineteenth century Scottish farmers and Scottish gardeners enjoyed a world-wide reputation.

Dumbarton is not a noted farming county. A considerable proportion of its surface is hilly, and therefore not well suited for agriculture. In this respect, however, Dumbarton is not so badly off as many Scottish counties. Only about a quarter of the total area of Scotland is cultivated, while about one-third of Dumbarton is either arable land or permanent pasture. The permanent grass lands are almost equal in area to the arable parts of the shire. The uncultivated parts consist mainly of moorland, which can be used as rough hill-pasture for sheep or can be preserved and let as grouse-moors.

The principal grain crop of the county is oats, with an area of nearly 7000 acres. This crop is much more suited to the moist, cool climate of the west of Scotland than wheat or barley, both of which require a sunny, dry, warm summer to thrive well. This explains why Fife and the Lothians can grow large quantities of wheat, while in Dumbarton the area under this grain is less than 500 acres, that is, less than one-twelfth part of the area devoted to oats. Hardly any barley is grown in the county. In 1910 there were only sixty acres of this crop. Potatoes claim a fair proportion of the arable land, there being over 2000 acres planted with them. Turnips are grown on 1300 acres. A large proportion of the cultivated ground is used for hay, which occupies over 10,000 acres—a greater area than any other crop.

Dumbarton compares favourably with Scotland as a whole with regard to the number of cattle and sheep it contains. In all Scotland there are about 63 cattle per thousand acres, whereas Dumbarton has 86. The county would be much more suitable for this branch of agriculture if only the lowlands were more extensive. The cattle are kept chiefly for dairy purposes, and therefore the great majority of them are Ayrshires. This breed has been found particularly suitable to the moist climate of the south-western counties. It is not only hardy, but also yields a larger proportion of milk to food consumed than any other breed in the country.

For the whole of Scotland the average number of sheep is 360 per thousand acres, while Dumbarton possesses over 400. The stock consists mainly of Cheviots and black-faced sheep. The wool of the black-face does not bring so high a price as that of the Cheviot, but the former breed is hardier and more suited to hilly districts.

12. Industries and Manufactures.

The manufactures of the Clyde valley began to develop markedly about the middle of the eighteenth century. It was not until then that the coalfields of the district could be properly worked, for coal could not be obtained in large quantities until James Watt had improved the steam engine. Then came the application of steam to manufactures, and the textile trade of Scotland began to thrive. The inventions of Hargreaves, Arkwright, Crompton, and Cartwright in England were of world-wide importance, and the Clyde valley was not slow in making use of the opportunity of creating new industries. One trade reacted on and stimulated another until the whole district became the hive of varied industries that we know to-day.

Dumbarton shared in the general prosperity of the Clyde valley. Indeed the growth of population in certain parts of the county has been phenomenal, and still continues. There is a wonderful variety, too, of industries and manufactures. Shipbuilding, marine-engineering, general engineering, dyeing and bleaching, coal-mining, chemical manufacturing, motor-car building, the making of sewing-machines—these are the most important industries, and it is obvious from this list that there is little specialisation. This is fortunate for the district as a whole, for it is seldom that several of the chief industries are notably depressed at the same time. Thus sudden fluctuations from excessive prosperity to the depths of adversity are not nearly so common in Dumbartonshire as in many other manufacturing districts. The same, indeed, may be said of the Clyde valley as a whole.

The principal industries of the county are shipbuilding and its ally, marine-engineering. Of the three shipbuilding counties of the Clyde, Renfrew, Dumbarton, and Lanark, the first generally produces somewhat more than a third of the total tonnage, and the other two are each responsible for a little less than a third. The Clyde is the greatest shipbuilding district in the world. In 1913 it built more than double the tonnage of any other district in the British Isles, except the Tyne. Of foreign countries, Germany ranked first for shipbuilding, yet in the year mentioned its total tonnage was surpassed by the single district of the Clyde by no less than 100,000 tons. In marine-engineering the supremacy of the Clyde is even more marked. The horse-power produced on its banks was greater in 1913 than that from any two foreign countries, or from the Tyne, the Wear, the Tees, and the Hartlepools put together. The Clyde designers, too, have always been in the van of progress. This is well illustrated by the

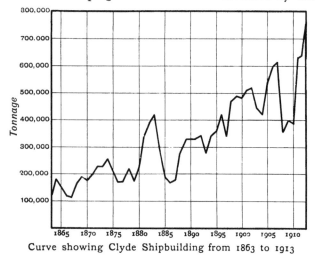

Curve showing Clyde Shipbuilding from 1863 to 1913

simple statement that the first passenger steamer to be built (the *Comet*), the first turbine passenger boat (the *King Edward*), and the first turbine ocean liner (the *Virginian*) were all constructed on the Clyde.

It is obvious that the Clyde possesses many weighty advantages for shipbuilding. The estuary runs into a busy coalfield where iron and steel working and marine-engineering are staple industries; and so material and skilled labour are close at hand. The human factor, too, has been no less important. The enterprise and foresight of the citizens of Glasgow in changing a shallow stream into a navigable river made the industry possible. The same diversity and variety mark the shipbuilding of the Clyde as we found characteristic of the industries as a whole. There is no specialisation on any one type of ship. From a Cunarder to a tramp, from a battle-ship to a motor boat, from an oil-king's steam-yacht to a dredger, from a white-winged racer to a square-rigged wind-jammer—every kind of craft that floats will be found on the stocks of the Clyde.

Of the Dumbartonshire shipbuilding firms three stand out pre-eminent, these three being John Brown and Company, Clydebank; William Beardmore and Company, Dalmuir; and William Denny and Brothers, Dumbarton. All these firms find a place in the first seven yards in 1913. In that year John Brown and Co. were second on the tonnage list, with the huge total of 82,722 tons. Of the five ships launched that year one was the mammoth Cunarder *Aquitania*, which just missed being the largest ship in the world by a few feet. A modern battleship is about 25,000 tons, but the *Aquitania* is nearly double that, namely, 47,000 tons. In marine-engineering, too, John Brown and Co. were *facile princeps* in 1913. The horse-power of the engines produced by that firm was the greatest ever made by any one firm in one year in the world's history.

Beardmore's firm is a remarkable one. Like John Brown and Co. they build a good deal for the Government, for the resources of their yard are equal to the largest battleship. This firm is probably unique in this country, for not only can it build a battleship, but can also protect it with armour plates, fit it with boilers and machinery, and even equip it with guns, all of its own

The *Aquitania*
(*Built by John Brown and Co.*)

manufacture. Denny's firm has always been very progressive. The yard is equipped with a large experimental tank, so that research and practice may go hand in hand. The first steamer in this country to be equipped with turbine engines was built in this yard. If the Clyde has specialised in any one type more than another, it may be considered to have done so in fast cross-channel passenger-boats.   This fact has a geographical basis. The beautiful estuary of the Clyde holds out so many attractions to town dwellers that it made a fleet of fast pleasure steamers an imperative necessity.   The type was therefore developed to a high degree of perfection, and so it is not surprising to find that Clyde-built boats, with few exceptions, monopolise the river and cross-channel traffic of the British Isles.   Denny's built the fast paddle and screw steamers for the Newhaven-Dieppe service.   Their most modern example of this type is the fine geared-turbine boat *Paris* for the same service, launched in 1913.   They built also the paddle-steamers for the Holyhead-Dublin service, for the Dover-Ostend, the Dover-Calais, and, coming nearer home, the splendid turbine-steamers for the Clyde estuary, the *King Edward* and the *Queen Alexandra*. John Brown and Company, too, have done notable things in the same line.   They built all the finest and fastest steamers for the Southampton-Channel Islands service and for the Southampton-Havre.   There are several other large shipbuilding firms in the county, but space prevents us from giving details regarding them. One minor aspect, however, of this industry deserves brief mention.   One or two places in the county have developed in recent years a flourishing business in building motor-boats.   Over thirty boats of this type were built in 1913 in Dumbarton, Rosneath, and Clynder.

In Dumbartonshire the chief branch of the textile trade is the bleaching, dyeing, and printing of cotton fabrics.   Calico-printing began in the Clyde valley in 1738, many years before it was introduced into Lancashire, where it is now so important. Bleaching with chlorine preparations was tried first in 1787 on the suggestion of that versatile genius, James Watt. Turkey-red dyeing is said to have been first introduced to the Clyde valley by a dyer from Normandy. This

Works of the Singer Company, Kilbowie

industry was revolutionised by the discovery of aniline dyes. Formerly it was an art that needed a lifetime for its mastery. The use of coal-tar products simplified and cheapened the manufacturing processes enormously. Increased brilliancy was obtained but at the cost of lasting qualities. The chief seat of calico-printing, bleaching, and dyeing in Dumbartonshire, and indeed in Scotland, is the Vale of Leven. Alexandria, Renton,

H.M.S. Ramillies

(Super Dreadnought of latest type. Displacement 29,350 tons. H.P. 40,000. Built by Wm Beardmore and Co.)

Bonhill, and Jamestown are engaged mainly in these trades. The cloth, however, is not made locally, but comes chiefly from England to be treated.

One of the principal industries of the county is the making of sewing-machines. The huge works of the Singer Manufacturing Company are situated at Kilbowie, and along with the shipyards have created a town of forty thousand inhabitants. When the enormous business of the company completely outgrew their works in Glasgow, it was intended to build a factory in Dumbarton; but the authorities of that town, with curious lack of foresight, put obstacles in the way. The Singer Company's works, therefore, were built at Kilbowie, and have kept on growing until they now occupy over fifty acres of ground. There are nearly 10,000 employees, and the net result of their labours is that every ten seconds of the working day a complete sewing-machine is created from its elements. When one remembers that this goes on hour after hour, day after day, and year after year, one is surprised that every family in Europe is not supplied with a machine. It is rather a coincidence that the greatest sewing-machine factory in Europe should stand opposite the greatest sewing-cotton factories, for Paisley is only four miles away across the river.

The other industries of the county are not specially distinctive. There is a fair amount of general engineering. One firm in Kilbowie has made a wide reputation for laundry machinery. Boilers, engines, tanks, castings, and countless other articles are also manufactured.

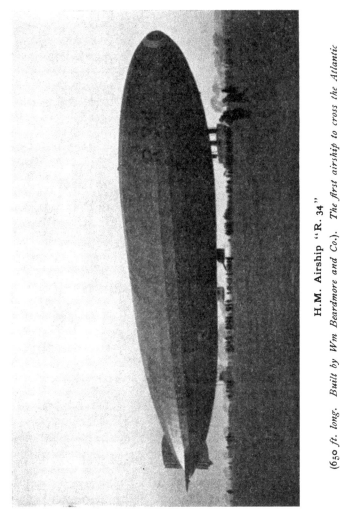

H.M. Airship "R. 34"

(650 ft. long.   Built by Wm Beardmore and Co.).   The first airship to cross the Atlantic

The foregoing account applies of course to normal pre-war conditions. During the war the balance of industry altered completely. This was particularly the case in Dumbartonshire and along the banks of Clyde, where everything was subordinated to the task of winning the war. Battleships were built instead of merchant-vessels, submarines instead of steam-yachts, "tanks" took the place of machine-tools, shells of sewing-machines, aeroplanes of motor-cars. The work of Wm Beardmore and Co. may be taken as typical of what was going on throughout the county. During the war this firm produced millions of shells, thousands of big guns, hundreds of aeroplanes from 80 h.p. Baby Sopwiths to giant 2000 h.p. Handley-Pages, scores of "tanks," and over 70 naval vessels of various types from battleships to submarines. Some characteristic examples of Beardmore's war-work are illustrated on pp. 95, 97, 99.

13. Mines and Minerals.

Dumbarton is not one of the most important mining counties of Scotland. It produces a fair amount of coal, considerably more than Renfrew, but not nearly so much as Lanark or Ayr. The explanation of Dumbarton's secondary position as a coal producer is that the true Coal Measures are practically non-existent in the county. The coal-seams that are worked occur in the Carboniferous Limestone series, and have been

Handley-Page Aeroplane

(4 *Engines.* 2000 H.P. *Wing spread* 126 *ft. Built by Wm Beardmore and Co.*)

worked for many years, a fact that is disagreeably emphasised by the number of "byngs" or waste rock heaps in the south-east of the county. Ironstone is also found in the same series, and is still mined, although the output has largely fallen off owing to the ore becoming worked out.

Over 2000 people are employed in coal-mines in the county, and in 1912 they produced 484,000 tons of coal. This seems a large amount at first sight until we learn that in the same period Lanark produced more than thirty-four times that total. The coal is extracted either on the "stoop and room" or on the "long wall" system. The former is now nearly extinct. On this system roads are driven through the coal and connected by cross passages, pillars of coal being left to support the roof. The roof is afterwards propped up by timber, and the coal pillars removed. The method was more applicable to the thick seams of a former day than now. On the long wall system, which is now generally followed, the whole of the coal is extracted at once. Beginning at the foot of the shaft, the "face" is gradually pushed outwards by the removal of the coal, while the waste material is stacked up to support the roof. This method is better adapted to the use of coal-cutting machinery, by means of which these seams can be made to pay, when it would be useless to work them by hand methods. The machines are often made to cut through the under clay so that all waste is prevented. Some of the machines are driven by electricity, and some by compressed air.

The amount of ironstone mined in Dumbartonshire

in 1912 was 17,000 tons, rather a small total when compared with the Ayrshire output of 217,000 tons.

A fair quantity of building stone is produced in the county. In the south-east the yellow sandstones of the Carboniferous Limestone series make an excellent building stone. The red rocks of the Old Red Sandstone series are quarried near Dumbarton and in the Vale of Leven. Most of the red building stone of this district is obtained from local quarries. On the whole this rock is not of first-class quality. It often contains impurities which are apt to produce ugly stains after the rock has been exposed to the weather. Some of the limestone bands of the Carboniferous system are of some industrial importance. Quite a thick band of the well-known Hurlet limestone outcrops near Duntocher, and is worked in places. The volcanic rocks of the county are of greater economic value. The Kilpatrick Hills and the numerous dykes and necks of volcanic origin offer an inexhaustible supply of road-metal. The ideal road-metal should be hard, tough, and not given to mud-making, offer a good footing in all weathers, and bind well together. The igneous rocks of the county possess all these qualities in a high degree. In its output of road-metal Dumbarton occupies quite a respectable position among the counties of Scotland. From the flat lands near the Clyde and in the Kelvin valley clay is obtained for brick-making.

14. Shipping and Trade.

The towns on the Clyde are fortunate in possessing a particularly favourable position for trade. By far the greatest commercial route in the world is that from Great Britain and the English Channel to North America. The Atlantic Ocean is now the chief highway of the world's commerce, and ports on the west side of Britain are therefore very favourably situated. The only large river estuary on the west coast of Scotland is the Clyde, and so it has practically a monopoly of west-coast shipping. Again, the estuary is well sheltered, and leads to the heart of the Central Lowlands of Scotland, by far the richest and most populous part of the country. Not the least of the advantages of the Clyde is that the rich coalfield of Lanarkshire stretches down the river as far as Glasgow, the head of ocean navigation.

From the earliest times, therefore, we find that the coasts of Dumbarton have possessed important centres of shipping and trade. The record goes back to pre-historic times. Dumbarton may even boast of possessing the earliest dock in the British Isles, for a part of the Dumbuck Crannog has been identified as a dock.

During the Roman occupation of Scotland there was an important naval station, Theodosia, on the site of Dumbarton. Coming now to mediaeval times, we find that in the thirteenth century Dumbarton was engaged in a dispute with Glasgow that lasted for centuries. Dumbarton claimed the right to tax all merchants from Glasgow trading with Lennox or Argyll. The quarrel became so bitter that the king, Alexander III, was forced to intervene. King James IV made Dumbarton his headquarters in his expeditions against the Western Isles in 1494 and 1495. Items in the Treasurer's books show that at that time Dumbarton was engaged in shipping and shipbuilding. One item is as follows: "To the byggin of the King's rowbarges bygite in Dumbartane, the tymmyre fra Loch Lomond and divers uthir woddis." While James was supervising the building of his ships on the Leven, he found time to hunt in the neighbouring woods, and at night amused himself at the "cartes" or listened to "evinsang." It was from Dumbarton that the child-queen, Mary Stuart, embarked for France in 1548.

In 1658 the magistrates of Dumbarton had the opportunity of greatly increasing the importance of their town by allowing it to be made the official port of Glasgow; but Glasgow's proposal was rejected because "the influx of mariners would tend to raise the price of butter and eggs to the inhabitants." In the eighteenth and nineteenth centuries the deepening of the Clyde adversely affected Dumbarton and other estuary towns. The shipping goes direct to Glasgow, the great trade centre. Dumbarton's shipping trade is therefore not of great importance. Nevertheless a large dock has been built and the Leven deepened. A fine pier has been made from the Castle Rock for river traffic. A depth of ten feet at low water is obtained at the pier.

Dumbarton has always been the principal centre of

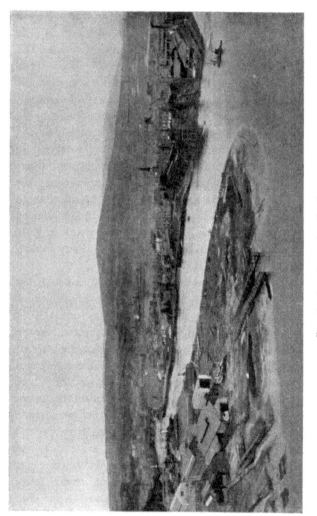

Dumbarton from the Castle

the shipping trade of the county, but one or two other places deserve brief mention. In the eighteenth century a good deal of smuggling went on along the coasts of the county, Row being particularly notorious. It carried on a large clandestine trade in whisky. Sir Walter Scott refers to the smugglers of Row in his *Heart of Midlothian*. When King George IV visited Scotland, he expressed a desire to taste genuine smuggled whisky. The Duke of Argyll, therefore, personally met the smugglers at Row Point and procured a barrel.

The harbour of the old-fashioned little town of Bowling has always a few ships of an unpretending kind. A fair amount of painting, repairing, and fitting out is done here. Quite a number of yachts and river steamers come to Bowling in the winter and spring for a general overhaul before the arduous labours of the summer commence. Bowling, also, is the western termination of the Forth and Clyde canal.

Most of the river steamers are now owned by the three great railway companies of the West of Scotland. The headquarters of the North British boats are at Craigendoran in Dumbartonshire. From this port there plies a fleet of powerful and speedy steamers, dashing back and forward between all the little pleasure towns of the firth. These dainty craft with their distinctive colour note in hull and smoke-stack add much to the charm and interest of the scenery of the Firth of Clyde. Craigendoran, Greenock, and Gourock, the three termini of the railway boats, have filched from the Glasgow Broomielaw much of its ancient glory.

15. The History of the County.

Authentic history begins in Scotland with the coming of the Romans. Julius Agricola, the Roman governor of South Britain, invaded Scotland and subdued the country as far north as the Firths of Clyde and Forth. In A.D. 81 he constructed a chain of forts between these two estuaries. One of the forts was built on Bar Hill, Dumbartonshire. Soon, however, the Roman troops were withdrawn from the forts, and Scotland was abandoned. Sixty years later southern Scotland was again occupied by the Romans, and this time their grasp was not relaxed until two and a half centuries had passed. In A.D. 140 a rampart was constructed along the line of Agricola's forts.

Little is known for certain of the history of Dumbarton, or of Strathclyde as a whole, for the next few centuries. The men of Strathclyde were Brythonic Celts, and represented the race which had occupied the southern part of modern Scotland at the coming of the Romans. The English invaders had pushed these Britons westward till their home was along the west coast from the Firth of Clyde to the English Channel. Not even in the face of the English foe did the Britons cease their quarrels; and after the battle of Arderydd in 573 the northern half of them transferred their chief seat from Carlisle to the rock of Alclwyd, afterwards Dumbarton; and at a later date the kingdom of Strathclyde stretched from Dumbarton to the Derwent in

Cumberland, but not including Wigtown and Kirk-cudbright. About this time some of the early Christian saints were connected with Dumbartonshire. St Patrick is reputed to have been born there, either at Old Kil-patrick, or Dumbarton, or Rosneath. He was carried off to Ireland in a raid made by the Scots and sold as a slave. After his liberation he made frequent journeys between Strathclyde and Ireland. St Kentigern, known as Mungo, *the beloved*, laboured first in Alclwyd. Owing to the jealousy of the King of Strathclyde he had to flee to Wales, but was recalled by the king's successor.

For over three hundred years the British kingdom of Strathclyde flourished, although exposed to the repeated and furious attacks of Picts, Scots, Angles, and Danes. The fortress of Alclwyd was more than once captured by the invaders and recaptured by the Britons. At the beginning of the tenth century the race of independent kings of Strathclyde died out. A brother of the King of Alban was then selected ruler by the Britons, and Strathclyde became dependent on Alban. Still the district remained a separate kingdom until the accession of Duncan in 1034, who became King of Scotia, with which Strathclyde was now incorporated.

The county is associated with the two great Scottish patriots. Wallace is said to have taken refuge from the English in the mountainous fastnesses of the north of Dumbarton. Near Rosneath, Wallace's Leap is still to be seen. Blind Harry says that the Scottish champion sacked the town of Dumbarton, which was in the hands of the English, burned the Castle of Rosneath, and was welcomed at Faslane on the Gare Loch by the Earl of Lennox. In 1305 all Scotland seemed again at the feet of Edward I. At this time the Governor of Dumbarton and Sheriff of the county was Sir John Menteith. He made Wallace prisoner near Glasgow, and very probably kept him for a few days in Dumbarton Castle. For about six centuries Wallace's great two-handed sword was kept in the castle until it was removed in 1888 to the Wallace Monument, at Stirling. Bruce sojourned for a time on Loch Lomond side during his wanderings; and Barbour draws a moving picture of his meeting with the Earl of Lennox, who rushed into his master's arms and wept aloud. The Bruce spent the evening of his days in Dumbartonshire, fishing, hunting, building ships, and sailing on the estuary. He died at Cardross Castle in 1329.

Other royal visits to Dumbarton have been already mentioned, and we may pass to the troublous times of James VI's minority. After Langside, Dumbarton Castle, to which Mary had wished to go, was held by the Queen's men for three years, and the story of its capture (2nd April, 1571) by the Regent's party is a thrilling one. Captain Craufurd of Jordanhill was the leader of the enterprise, and he had the help of a man who had once been warden of the castle, and knew every foot of the rock. Craufurd and a hundred men set out from Glasgow after sunset, and by midnight had almost reached the town. Ropes and scaling ladders were now served out, and the party advanced to the rock. The ladders were placed in position and the soldiers began to mount.

The ladders slipped, however, and made so much noise that, if a proper watch had been kept, the assailants must have been discovered. Another attempt was made, and by toiling hard with ropes and ladders the middle of the rock was scaled as dawn was breaking. The ladders were planted again, and now an unlooked-for obstacle was met. One of the men while climbing was seized with a fit, and grasped the ladder so convulsively that his hands could not be loosened. What was to be done? He could neither be passed nor his clasp slackened. Craufurd's fertile brain devised a plan. The soldier was tied to the ladder, and the ladder turned round, leaving the way clear. Soon the wall of the castle was reached; and, hooking their ladders once more, the assailants arrived at the top. The alarm was given, but Crauford and his men dashed upon the unsuspecting garrison, who offered a feeble resistance, and the castle was soon in the power of the Regent. So bold and unexpected was the attack that Crauford did not lose a single man.

In the time of the religious persecutions we hear of the castle being taken by the Covenanters and recaptured by the Royalists. Since that time Dumbarton Castle has suffered no assaults except by the silent hand of time.

Glen Fruin was in 1603 the scene of a desperate battle between the Clan Colquhoun and the Clan Macgregor. The Macgregors had left their own side of Loch Lomond and made a raid into the country of the Colquhouns. With "aixes, twa-handit swordis, bowis and arrowis, and with hagbutis and pistoletis," they marched into Luss territory. The Colquhouns were enticed into Glen Fruin by a party of Macgregors who waited for them at the head of the glen. Another detachment of the raiders cut off the retreat of the Luss men at the foot of the glen. The Colquhouns were slaughtered without mercy. Then the houses were

The "Highlandman's Road" near Helensburgh

fired and the cattle carried off. By this dreadful deed the Macgregors drew on themselves the wrath of the government. Thirty-five of them were executed, and parliament enacted that the very name of Macgregor should no longer exist in Scotland. Death was the fate of anyone bearing that name or any one found giving food or shelter to a Macgregor.

By the beginning of the eighteenth century the Macgregors must have regained strength to no small degree, for we find them threatening to take an active part in the Jacobite rebellion of 1715. They seized all the boats on Loch Lomond and collected them at Rowardennan ready for a sudden fray. Word of this reached the men of the western lowlands, no friends of the Stuarts. An armed expedition, therefore, went up the River Leven in boats, and on reaching Loch Lomond, sailed for the opposite shore to meet the Macgregors. As they crossed the loch they fired their guns and small arms, and altogether made so terrible a display that when they reached the eastern shore no Highlanders were to be seen. The boats, however, were discovered. Some were destroyed and some taken away; and in this way further trouble from raiding Macgregors was obviated.

In recent years the history of the county has been a peaceful record of rapid growth in population and material wealth, of progress in commerce and industry, in scientific discoveries, particularly those applied to industries, and, in general, of advancement socially and materially.

16. Antiquities.

By far the most important relic of antiquity in the county is the Roman Wall[1]. In the time of the Emperor Antoninus Pius a continuous rampart—generally called

---

[1] The paragraphs on the Roman Wall are based mainly on Dr Macdonald's work *The Roman Wall in Scotland*.

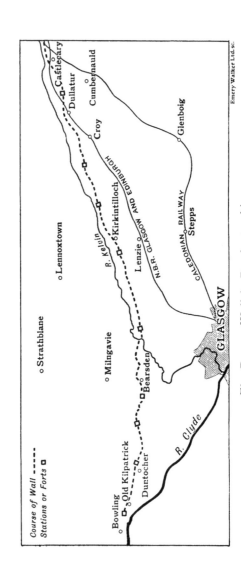

Course of Wall - - - -
Stations or Forts ▫

The Roman Wall in Dumbartonshire

(*After Dr Macdonald*)

Emery Walker Ltd. sc.

Antonine's Wall—was erected along the line of Agricola's forts between the Clyde and the Forth, for a distance of thirty-six miles. The work was entrusted to Lollius Urbicus, governor of Britain, A.D. 140 to 143.

The barrier was two-fold, a rampart and an outer ditch. About forty or fifty yards south of the rampart ran the military road right across Scotland. Forts were built along the wall at irregular intervals of about two miles. The rampart is now mainly levelled to the ground. Where it remains, it is usually about two or three feet high. The rampart was not a wall such as we are accustomed to see. It was about fourteen feet broad at the base and roughly six feet broad at the top. Its height was originally about ten feet. The foundation was built of stone. The rest of the rampart was constructed of sods, course upon course, like a wall of stone. The external ditch was about forty feet wide, except where it had to be cut out of solid rock, when naturally it was narrower. It was perhaps about twelve feet deep on an average, in places even deeper. The military road was made of stone. A foundation of large stones was laid and covered with smaller ones. It was cambered for drainage just like a modern road.

The wall begins near Old Kilpatrick church: Chapel Hill may be taken as the western termination. It can be traced through Duntocher and then to Bearsden. It crosses to the south side of the river Kelvin near Summerston railway station. Keeping now roughly parallel to the Kelvin, it goes through Kirkintilloch and then to Dullatur. It finally passes out of Dumbarton almost at Castlecary station. Several of the forts in the Dumbarton portion of the wall have been carefully investigated. There are fairly clear remains of a fort at Duntocher. In it a gold coin of the time of Hadrian was found, and several other coins in the vicinity. The most interesting find was the statuette of a woman, now in the Hunterian Museum, Glasgow University. Near the fort the remains of baths have been discovered.

Inscribed Tablet from the Roman Wall

The fort itself was a fair-sized structure, being three or four hundred feet long. From a fort at Auchendavy a rich collection of Roman remains was secured. They included coins, pottery, altars, a mutilated bust, and a small intaglio of lapis-lazuli, which had once been in a signet ring. Excavations at Bar Hill afforded much new information regaiding the forts. From a well over twenty pillars were taken, in addition to a large altar and inscribed stones. In other parts of the fort were found pottery, pieces of glass bottles, rude sculptures, querns, shoes of men, women, and children, money, bones of oxen, and shells of mussels and oysters.

The inscribed stones found at various places along the course of the wall are of special value, for they give us exact information on several points of the highest importance. They comprise commemorative tablets, altars, and tombstones. The commemorative tablets are of great interest. They record that a particular body of soldiers had executed a certain piece of work, in two cases the work being definitely specified as "opus valli," the work of the rampart. These tablets are unique, for nothing similar has been discovered from Roman remains elsewhere. Fifteen tablets out of the total of seventeen discovered may be inspected in the Hunterian Museum, Glasgow. The accompanying photographs show three of them, all found in Dumbartonshire. The figure on p. 114 represents a plain slab that came from near the western termination of the rampart. It is one of the two slabs found that definitely state the work to have been the construction of the wall. The inscription is:

IMP · C · T · AELIO ·
HADRIANO · ANTO
NINO · AUG · P · P ·
VEX · LEG · VI · VIC ·
P · F · OPVS · VALLI
P · ∞ ∞ ∞ ∞ C · XLI

Expanded this would become "Imperatori Caesari Tito Aelio Hadriano Antonino Augusto Patri Patriae

Vexillatio Legionis Sextae Victricis Piae Fidelis Opus
Valli Perfecit MMMMCXLI." A literal translation is as
follows: "In honour of the Emperor Caesar Titus Aelius
Hadrianus Antoninus Augustus, Father of his Country,
a detachment of the Sixth Legion, the Victorious, the
Dutiful, the Loyal, completed of the rampart 4141 feet."

Inscribed Tablet from the Roman Wall

The illustration on p. 116 is a highly ornamental
tablet. The figure represents a Victory holding a palm-
leaf and resting on a globe. The inscription is: "In
honour of the Emperor Caesar Titus Aelius Hadrianus
Antoninus Augustus Pius, Father of his Country, a
detachment of the Twentieth Legion, the Valerian, the
Victorious, completed 4411 feet." This slab came from
the western portion of the wall, but the exact spot is
not now known. The tablet shown in the figure on p. 117
also came from Dumbarton, probably the parish of Old

Kilpatrick. The tablets vary in size from less than three feet to over nine feet in length. They were probably brightly coloured when first set up.

Early in the reign of Commodus (A.D. 180–192) the rampart was finally abandoned by the Roman troops. There is some evidence for thinking that the evacuation was deliberate and orderly. The stores were burned, the altars and tablets were thrown into wells or buried

Portion of Inscribed Tablet from the Roman Wall

in the earth, the buildings were demolished; and then the legionaries turned their faces southward.

Some years ago a remarkable pile-structure or "crannog" was discovered near Dumbuck. The structure itself is a ring of oak piles four feet above ordinary low water, and five feet below high tide level. There are twenty-seven piles in all, forming a circle fifty feet in diameter. Within this circle there is a kind of flooring of horizontal timbers. From the centre of the structure a pavement of stones extends for twenty yards till it meets a breakwater that stretched completely round the structure. A short distance away an oak canoe was

found containing ornaments and tools of various kinds.
The canoe was thirty-six feet long, one of the largest
ever found in the Clyde valley. When discovered, it
was lying in a peculiar dock-like structure, which was
connected with the pile building by a causeway of timber
and stone. A large number of small objects were found
in the immediate vicinity of the pile-structure, comprising
bones of animals and birds, flint scrapers, hammer
stones, pointed bone tools, stone ornaments, spear heads
of slate, and figures with grotesque human faces carved
from shale. No pottery or metals were found. Similar
objects were found not far away at an old fort on
Dumbowie Hill. Doubts have been expressed regarding
the genuineness of some of these finds, because they are
unknown from other prehistoric pile-structures.

The other relics of antiquity found in the county are
of minor importance. There are several stones to
which tradition has attached associations with the early
saints. A carved stone in Old Kilpatrick churchyard
is supposed to represent St Patrick; but modern
sceptics find a greater resemblance to an armed knight,
perhaps one of the powerful Colquhoun family. St
Kessog, an Irish missionary of the sixth century, is said
to have been martyred near Luss. The spot is marked
by a cairn called Carn-na-Cheasog or Kessog's Cairn.
An effigy on the cairn was removed to the chapel at
Rossdhu. The effigy is certainly not of the sixth
century, but may be a thirteenth or fourteenth century
representation of the saint. In the grounds of Mount
Blow, a cross was found similar to the well-known cross

at Barochan, Renfrewshire. For many years it was used as a foot-bridge and suffered considerable injury. In 1891 a broadsword was discovered in a stream at Auchentorlie. It has an open, steel-work hilt of basket pattern. On one side was the word "Andrea" and on the other a defaced inscription which was most likely "Ferara." There seems little doubt that it is an example of the work of the famous sixteenth-century sword-maker of Italy.

17. Architecture—(a) Ecclesiastical.

Well-preserved specimens of ancient architecture are very rare in Dumbarton. In this respect it is far poorer than any of the other shires that form the Clyde valley. Lanark, Ayr, and Renfrew offer incomparably more fruitful fields to the student of early architecture

The scanty ruins of one of the oldest churches in the county are in the town of Dumbarton. The Collegiate Church of St Patrick was founded in 1450 by Isabella, Duchess of Albany and Countess of Lennox. It was subsidiary to the Abbey of Kilwinning. After the Reformation the church was allowed to fall into ruins, and much of the material was carried off for building purposes. A single tower arch is all that is left, and the arch is not even in its original position, for in 1850 it was removed to make way for a railway station. The old parish church was built originally about 1565, but no trace of it now remains. A drawing of it made in

1747 shows a quaint, cruciform building with a spire, a high roof, and crow-stepped gables. The old church was demolished in 1810. Near Cardross at Kirkton of Kilmahew a chapel existed in 1370, for it is mentioned in a charter. In 1467 a new chapel was built, and the present chancel was probably a part of this building. The gables are crow-stepped, and there are shields with the sacred monogram I. H. S.

The Reformation put an end to mediaeval, ecclesiastical architecture in Scotland. A few churches were certainly erected under the influence of the Episcopalians, but the Presbyterians attempted to eliminate everything that savoured of the old forms, and to this end were content to erect buildings that had absolutely no claim to respect so far as their architecture was concerned. In the eighteenth century, however, there arose in England a distinct revival of the interest in architecture, and particularly in classical styles. This feeling hardly stirred in Scotland till the nineteenth century. We are told that in the eighteenth century the Scottish churches "were disgraces to art and scandals to religion. They were mean, incommodious and comfortless; the earth of the graveyard often rose high above the floor of the church, so that the people required to descend several steps as to a cellar, before they got entrance by stooping into the dark, dismal, damp and hideous sanctuaries." At the beginning of the nineteenth century, however, a great change for the better began to take place. Architects made a special study of old buildings and old styles; and this combined with the rapidly increasing wealth of the country was soon reflected in many noble ecclesiastical buildings. The great and wealthy industrial communities of Dumbartonshire can now without exception boast of modern churches that will bear comparison with those of mediaeval times.

18. Architecture—(*b*) Military and Domestic.

The finest of the old castles of Scotland were erected in the thirteenth century. The nobles were rich, labour was cheap, and the feud with England had not yet become chronic. It is to this period that the magnificent pile of Bothwell Castle belongs. The end of the thirteenth century, however, marks a great change in the style of the castles of Scotland. The War of Independence completely exhausted the resources of the country, and consequently we find that large and massive buildings, such as Bothwell Castle, were no longer erected. Their place was taken by strong, square towers, simply fashioned after the model of the Norman keeps. These are the castles of the second period. They are specially characteristic of the fourteenth century, but continued to be built at much later dates; and from the simplicity of the design it is often difficult to determine the exact age. In the fifteenth century the plan was slightly elaborated. The simple, square, keep-like style was retained, but the castle was built round a central quadrangle or courtyard. In addition a separate tower or keep is often found, capable of being defended although the rest of the castle should be captured. These are the third-period castles, and most of the Dumbartonshire structures are examples of this class. All the castles built in the reign of James I and until that of James V are of the modified keep style. All this time, however, the defensive features were becoming less in evidence, while domestic requirements were demanding more consideration. Thus we find that what was originally a necessary feature of successful defence became later merely ornamental; and, while the thick walls were retained, they were honey-combed with chambers.

Of the castles of the first and the second period no good examples exist in Dumbarton. There are one or two relics of the simple keeps of the third period. Badenheath Castle in Kirkintilloch parish is a fine example of a peel, dating from the end of the fifteenth century. The keep was oblong in shape but only a half remains. It was built of fine, coursed masonry, which proved an irresistible temptation to builders of a later age. The design of the entrance doorway is remarkable. A fine hall with a mantelpiece is also noteworthy. Fragmentary remains of another third-period castle may be seen at Dunglass, near Bowling. It also has been much plundered by later builders. Even the Commissioners of Supply, who should have known better, in 1735 ordered it to be used as a quarry for material to build a new quay. The old building is now inset in modern work. The round tower seems to be a seventeenth-century structure. Dunglass Castle was one of the strongholds of the Colquhouns.

With the end of the third period there is a marked break in the continuity of the style of architecture. Gothic severity is often replaced by florid ornament of Renaissance style. The change was facilitated by the Union of England and Scotland in 1603. An impregnable stronghold was no longer either desirable or possible, for organised lawlessness had largely disappeared, and thick stone walls were no defence against artillery. Thus we find that, while the plan of the buildings remained the same, the external appearance and the details were altered. The transition from military to domestic needs is shown in such a change as the evolution of angle turrets into bow windows, a change that is typical of the whole process. Kilmahew Castle, near Cardross, is an example of the fourth-period castles. It was of oblong shape, but only two sides are left. A corbelled parapet surmounted the walls. The castle was the ancient seat of the Napiers. Bannachra Castle, in Glen Fruin, also belongs to the fourth period. Three sides of the original oblong are left. The gables were crow-stepped. Darleith Castle is an old keep now incorporated in a modern mansion, three miles north of Cardross. A dormer window and some coats of arms are well preserved. Only one wall is left of Rossdhu Castle. It was a simple, square keep of the fourth period. The castle was occupied till 1770, when it was partially demolished to supply materials for a modern mansion.

Although poor in historic ruins, Dumbarton possesses a large number of fine, modern mansions. The beauty of many parts of the county combined with its proximity to the large industrial centres of the Clyde valley contribute to the value of the residential estates. Only

Ardencaple near Helensburgh

a few of the better known seats of the county can be selected for mention. Ardencaple, a fine mansion not far from Helensburgh, is in the old Scottish baronial style, and is chiefly modern. Parts of it, however, are very ancient, and are said to date as far back as the twelfth century. It is one of the chief seats of the Colquhouns of Luss, but for a long time it was the residence of the Dowager Duchesses of Argyll. Garscadden, in New Kilpatrick parish, is a historic estate. It was held by the Flemings, the Erskines, and the Galbraiths; and about 1664 passed to the Campbell Colquhouns of Killermont. The mansion is noted for a remarkable, castellated Gothic gateway, the most imposing in the west of Scotland. A grim story is told

Rossdhu House near Luss

of one laird of Garscadden which illustrates forcibly the manners of a past generation of Scottish landed proprietors. Garscadden had been dining with some neighbouring lairds, and, as usual, the drinking had been long and deep. One of the company noticed an unusual expression on Garscadden's face, and remarked, "What gars Garscadden look sae gash?"[1] "He may

[1] Wise, sagacious.

weel look gash," replied another, "for he's been wi' his Maker these twa hours. I saw him step awa, but I didna like to disturb guid company."

Killermont House, on the banks of the Kelvin, is a spacious and well-proportioned mansion, built partly about 1805 and partly at an earlier period. The grounds are magnificently wooded. The house is now the clubhouse of the Glasgow Golf Club, while the estate has been transformed into a beautiful, inland golf course. Rossdhu, three miles south of Luss, is the principal seat of the Colquhouns of Luss. The estate belonged in former times to the Earls of Lennox, but came into the hands of the Colquhouns in the fourteenth century. The house, a handsome mansion built about 1774, was visited by Queen Victoria in 1875. The grounds are beautifully wooded.

19. Communications.

The existence of routes depends on the demands for communication between one part of a district and another; the nature and details of the route depend mainly on the physical features of the district. High ground is a barrier to communication, often so serious as to divert routes for many miles out of their shortest course. It is plain that in Dumbarton the problems of communication are quite different according as we are dealing with the mountainous north-west or the flat south-east.

East of the Kilpatrick Hills the county presents few

obstacles to communication; and roads and railways therefore go in a fairly straight line. The Kelvin valley offers the easiest route between the west and the east of Scotland. This valley is continuous with the Carron valley leading to the Firth of Forth; and by keeping to this Kelvin-Carron route one can cross Scotland completely without ever rising 200 feet above sea-level. It is for this reason that the Forth and Clyde canal takes this route. The main line of the North British Railway between Glasgow and Edinburgh runs parallel to the canal for many miles. A portion of the line is in Dumbarton. The railway enters the county at Lenzie and leaves it at Castlecary. This route is also followed by one very important main road, the highway between Glasgow and Stirling. This road enters the detached eastern part of Dumbarton at Molinburn and leaves it at Castlecary.

In the north-western part of the county the chief factors that control the routes are the Kilpatrick Hills and the Highland mountains in the peninsula almost encircled by the Gare Loch, Loch Lomond, and Loch Long. Railways and roads are forced to circumvent these obstacles, for they cannot surmount them. High ground is much more formidable to railways than to roads. While a gradient of one in forty is a stiff incline for a railway, there are roads with gradients as high as one in five, although this is excessively steep. We find, therefore, that the railways are confined to the coastal strip or very low valleys, while occasionally a secondary road may cross the lower parts of the hills.

The North British Railway runs from Glasgow through Dumbarton to Helensburgh, for the whole of its course keeping to the low shelf furnished by the raised beaches. At Craigendoran a branch runs to the piers

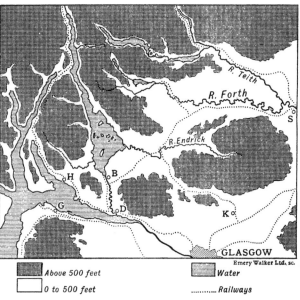

Above 500 feet   *Water*

0 to 500 feet   .... *Railways*

Emery Walker Ltd. sc.

Sketch Map showing how communications in Dumbartonshire are affected by high ground

that form the headquarters of the smart N.B.R. river steamers. At Craigendoran, too, the West Highland Railway begins as a separate line. It goes along the eastern shore of the Gare Loch, and then crosses to Loch Long by a deep notch in the hills. The flat coastal strip along Loch Long is so narrow that there is room only for the road. The railway, therefore, is cut along the hill-side, in some places with a sheer drop from the line into the loch. At Arrochar the railway makes use of another of the curious passes in order to cross from Loch Long to Loch Lomond. The way is then fairly easy to Ardlui at the head of Loch Lomond. The line then runs up Glen Falloch until it passes out of the county into Perthshire. The scenery all along this route is very fine.

A branch of the Caledonian Railway goes from Glasgow to Dumbarton. From Dumbarton to Balloch the line is a joint one between the Caledonian and the North British Railways. In summer this is a busy tourist route, for steamers meet the trains at Balloch, and carry passengers to the head of Loch Lomond. A mile south of Balloch a line branches east through Jamestown, and skirts the northern edge of the Kilpatricks on its way to Buchlyvie and Aberfoyle. Another line worth notice is the suburban one from Glasgow to Milngavie.

In the lowland part of the county high-roads cross and recross each other in every direction. In the Highland part the roads keep mainly to the shores of loch or lake, or pass through deep valleys. The modern tendency is for traffic to keep lower and lower down. A number of the old roads of the county rise several hundred feet above sea-level; but, as traffic by road and rail becomes faster and faster, it is found that the lowest road is ever the shortest. Although communication in

Dumbartonshire has been kept up for many centuries along the routes indicated, yet proper roads are of comparatively recent origin. In former times wheeled traffic was hardly possible, and most of the trade was done by pack-horses. Before the Clyde was deepened so that boats could come to Glasgow, pack-horses carried the goods from Dumbarton or Port Glasgow. The passing of the Turnpike Roads Act in 1751 marked the beginning of a new era; and good roads gradually replaced the old horse tracks. A further improvement was heralded by the establishment of County Councils in 1889, and the transference to them of the care of the roads. In recent years the development of fast motor traffic has presented a new problem to road authorities; and it may be that we are but at the beginning of an altogether new phase of road construction rendered necessary by the modern craze for speed. A few years ago a national Road Board was organized, to which all taxes on motor-cars and petrol are paid. This Board is empowered to give grants to County Councils, not for upkeep, but for improvements of roads.

The Forth and Clyde Canal, opened for traffic in 1790, has its western portion in Dumbartonshire. The canal begins at Bowling and for some miles keeps parallel to the Clyde. At Clydebank it turns east, and near Maryhill passes for a time into Lanarkshire. It enters Dumbartonshire again at Kirkintilloch, and for several miles keeps close to the river Kelvin, finally leaving the county near Castlecary. The eastern termination of the canal is at Grangemouth. The maximum height above sea-level attained by the canal is 156 feet; and thirty-nine locks in all are needed to raise boats to this level and lower them again to the sea. In 1867 the canal was bought by the Caledonian Railway Company.

The canal is associated with some of the earliest and most important experiments in steam navigation. In 1789 Miller and Symington constructed a steamship which was tried on the canal, and attained a speed of six miles an hour. Symington continued his experiments, and in 1802 launched the famous *Charlotte Dundas*, a stern-wheel steamer, which towed two laden barges on the canal for a distance of twenty miles. This is generally agreed to have been the first quite successful experiment in steam navigation.

During the first half of the nineteenth century canals were of great importance for the carriage of both passengers and goods. From one station on the Forth and Clyde Canal at Kirkintilloch 20,000 passengers were carried every year. In 1832 five iron steamers with stern paddles plied on the canal. But the palmy days of canal traffic both for passengers and goods have passed away. As railways were extended the importance of canals declined. The complete explanation of this fact is by no means easy. It has been attributed to their passing into the control of railway companies, but this explanation is not satisfactory. The smallness of the vessels in use, and the consequent additional handling of goods, undoubtedly militate against the greater use of canals in these days, when the whole tendency is to

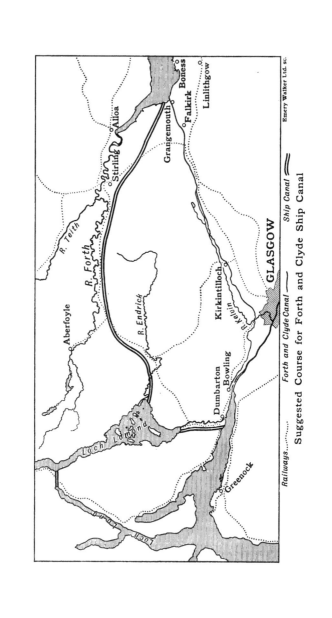

Railways........ Forth and Clyde Canal —— Ship Canal ⩳

Suggested Course for Forth and Clyde Ship Canal

Emery Walker Ltd. sc.

handle and carry goods in as large quantities as possible. With the adoption of improved methods of traction or propulsion, there seems no good reason why the importance of canal traffic should not to some extent be restored.

A proposal to construct a deep ship-canal between the Forth and the Clyde has recently been much discussed. The western portion of such a canal would certainly pass through Dumbartonshire. One scheme would make the canal take pretty much the line of the present Forth and Clyde Canal. On the other scheme the canal, starting from the Forth near Grangemouth, would pass south of Stirling and up the Forth valley for several miles. Coming round the north-west face of the Campsies, it would then meet the Endrick valley, and so pass to Loch Lomond. From Loch Lomond there is a choice of two exits, both in Dumbarton. One possible route is down the Leven to Dumbarton, the other is across the narrow neck between Loch Lomond and Loch Long. Undoubtedly great advantages, both economic and strategic, would follow the construction of such a waterway. The great drawback is the enormous cost, which necessitates government assistance. The cost has been roughly estimated at £20,000,000 for a depth of 36 feet.

20. Administration and Divisions.

At first the shire was administered by hereditary sheriffs, the office being generally held by one of the powerful families of the county. In 1747 hereditary jurisdiction was finally abolished; and since that time appointments have been made by the Crown. At the head of the county is the lord-lieutenant, selected generally from the local nobility or the land-owning class; but his position is mainly an honorary one. He is assisted by a vice-lieutenant and a large number of deputy-lieutenants and justices of the peace. The actual administration of the law is carried out chiefly by sheriff courts held at Dumbarton. These are presided over as a rule by the sheriff-substitute, for the sheriff-principal is generally a practising advocate. The police force is a county constabulary, except in the large burghs, which have their own separate forces. The county returns two members to parliament. One member is returned by the burgh of Dumbarton, and another by the rest of the county.

County Councils were established in 1889, and look after the finances, roads and bridges, water supply, public health, police, and general administration. The unit of poor-law organisation is the parish, and the poor-laws are administered by parish councils. There are twelve parishes in the county: Arrochar, Bonhill, Cardross, Cumbernauld, Dumbarton, Kilmaronock, Kirkintilloch, Luss, New Kilpatrick, Old Kilpatrick, Rosneath, and Row. There are in addition a number of burghs, largely independent of the county council: Clydebank, Cove and Kilcreggan, Dumbarton, Helensburgh, Kirkintilloch, and Milngavie. Dumbarton is the only royal burgh. The burghs are administered by town councils, which manage the property of the burghs, impose the rates necessary for upkeep, and make bye-laws for the regulation of the trade of the town and the conduct of the inhabitants. Town councillors are elected for three years, and one-third of the council retires annually. The councillors elect among themselves magistrates, who, besides performing other duties, act as judges in the cases that come before the ordinary police courts.

It must not be forgotten that there is still a considerable amount of overlapping and confusion in the administrative divisions, not only of Dumbarton but of all the counties of Scotland. The registration county is not the same as the civil county; the ecclesiastical parish differs from the civil parish; the district under municipal authority has no fixed relation to any of these other areas. In 1889 under a Local Government Act the Boundary Commissioners rectified some of the most glaring anomalies, and transferred areas from one parish to another, sometimes from one county to another. The ecclesiastical divisions, however, in many cases still fail to harmonise with the civil divisions.

Under the Education Act of 1872 the management of education in Scotland was entrusted mainly to School Boards, of which Dumbarton had fourteen. Education, free to all, was compulsory for children between the ages of five and fourteen years. Secondary and technical education was financed largely by a County Committee, which was empowered to give grants to schools, and to assist pupils by bursaries or otherwise. By the Education Act of 1918 School Boards and County Committees were abolished, and replaced by a single body, the

County Education Authority. The school leaving age was raised to fifteen years, and compulsory Continuation Classes were instituted until the age of eighteen years. Above the primary schools there are two classes of higher schools called Intermediate and Secondary. The former school provide a three years' course, and the latter at least a five years' course of education after the elementary stages. Pupils who have passed through a secondary school with credit are quite able to go with profit direct to the University. The pupil teacher system is to a great extent abolished. On passing the examination after three years in an intermediate school, young people who wish to become teachers may be accepted as junior students. They take the usual curriculum of a secondary school with some training in the art and science of teaching in addition. They then pass to the training college as students in full training, where they spend two years, after which they are recognised as certificated teachers, although two years of probation must be passed before final recognition is obtained. There are centres for the training of junior students at Bonhill, Dumbarton, Lenzie, Clydebank, and Helensburgh.

21. Roll of Honour.

It must be confessed that the Roll of Honour of Dumbartonshire is neither long nor specially distinguished. There is not one name of absolutely first rank identified with the county. Compared with Renfrew or Lanark or Ayr, the number of distinguished men born in the shire is surprisingly small.

Among the famous persons connected with Dumbar-

Tobias Smollett

ton we have already mentioned St Patrick and St Mungo, Wallace and Bruce, James IV and Mary Stuart. George Buchanan, who read Livy with Mary and afterwards became tutor to her son, James VI, was born in

Stirlingshire, but is said to have received part of his
education in Dumbarton. Buchanan was the greatest
scholar of his age.

Sir James Smollett of Bonhill, grandfather of the
novelist, was born in Dumbarton. He was trained as
a lawyer, and represented the county in parliament.
For his services to William of Orange in 1688 he was
knighted and made a judge. He was a strong supporter
of the Union of 1707. His grandson bears the most
distinguished name of any native of the county. Tobias
George Smollett was born in 1721 at Dalquhurn, near
Renton. The site of the house is now occupied by a
parish church. The novelist-to-be was educated at
Dumbarton Grammar School and early made a name
for himself for biting, satirical verse. Smollett was
trained as a surgeon, but never succeeded in practice.
The success of his novel *Roderick Random* (1748) turned
him altogether to writing. His last and best book is
The Expedition of Humphrey Clinker. A tall obelisk with
a Latin inscription was erected to his memory by a
cousin.

Henry Bell, the father of steam navigation in
Europe, was born in Linlithgow in 1767. When forty
years of age he removed to Helensburgh, where his wife
kept an inn, and he devoted himself to experiments in
engineering. Long before this time, however, he seems
to have conceived the design of applying the steam-
engine to the propulsion of ships. The famous *Comet*
was built to his designs, and the engine was made by
himself. He died at Helensburgh in 1830.

David Gray was born in 1838 on the banks of the
Luggie near Kirkintilloch.   His father was a hand-loom

The Engine of Bell's *Comet*

weaver, and David was the eldest of eight children.
At school he showed so much promise that he was
destined for the ministry, and proceeded to Glasgow
University. There he formed an intimate friendship

with Robert Buchanan, and the two determined to go
to London and embark on a literary career.   Gray

Principal Story

struggled for a time in London against poverty and
consumption, but returned at last to Kirkintilloch to die.

The young poet's pathetic career came to an untimely end in 1861. He was only twenty-three when he died. His longest poem, a description of the Luggie, a tributary of the Kelvin, contains many beautiful passages. His best work is contained in a series of sonnets written in the knowledge that death was not far away.

Rosneath parish is associated with one or two famous names. Professor Matthew Stewart, the mathematician, and father of Dugald Stewart, was for some time minister of the parish. John Anderson, the founder of Anderson's College, Glasgow, was the son of another minister. In more recent times Robert Story and his son Herbert Story, Principal of Glasgow University, were distinguished holders of the same office.

22. THE CHIEF TOWNS AND VILLAGES OF DUMBARTONSHIRE.

(The figures in brackets after the names of the burghs give the population in 1911, and those at the end of each section are references to pages in the text.)

Alexandria is situated on the right bank of the river Leven opposite the town of Bonhill, with which it is connected by a bridge erected by one of the Smolletts of Bonhill. Alexandria owes its rise to the establishment of the bleaching, printing, and dyeing industries. Its trade received an impetus from the establishment there of the works of the Argyll Motor Company. On the failure of that firm the works were taken over by the Armstrong-Whitworth Company. (82, 94.)

Arrochar (537) is a small village at the head of Loch Long. It is surrounded by deep glens and high mountains, and is therefore a favourite tourist resort. (20, 48, 49, 63, 65, 76, 129.)

Balloch, from the Gaelic *bealach,* a pass, is situated where the river Leven emerges from Loch Lomond. It is a pretty village much frequented by anglers and tourists. Near Balloch a considerable extent of ground bordering Loch Lomond has recently been acquired as a public park by the Corporation of Glasgow. (129.)

Bearsden (3257), five miles north-west of Glasgow, is a residential suburb of Glasgow, and consists mainly of well-built villas and cottages. A branch of the North British Railway connects it with the city. The pretty surroundings have led to the establishment of several large educational and charitable institutions in the neighbourhood of the village. (82, 113.)

Bonhill, on the left bank of the river Leven, stands opposite to Alexandria, and, like it, is a place of recent origin, owing its growth to the turkey-red works in the neighbourhood. The estate of Bonhill at the Restoration came into the hands of Sir James Smollett, grandfather of the novelist, and still remains in possession of that family. (96, 136, 138.)

Arrochar and the Head of Loch Long

Bowling (973) is a small seaport nearly four miles east of Dumbarton. Behind the village the Kilpatrick Hills rise steeply in a high escarpment. There is a shipbuilding yard near the harbour and a small distillery in the village. Near at hand is Dunglass Castle and the monument to Henry Bell. (55, 105, 122, 130.)

Cardross (1520) is a pretty village situated four miles north-west of Dumbarton. It is a residential place and a summer resort. (50, 57, 74, 78, 120, 123.)

Clydebank (37,547) is the largest town in the county. Its rise was due primarily to the settling there of two great industrial concerns, a large shipbuilding yard (now John Brown & Co.) and the huge factory of the Singer Sewing Machine Company. Other firms came to the locality, and now one town merges into another almost all the way from Glasgow to Bowling. During the ten years 1901—1911 the population of Clydebank increased by nearly 80 per cent., much the largest increase in Scotland. (26, 84, 91, 130, 134, 136.)

Cove, Loch Long

Clynder, a small village on the west side of the Gare Loch, has a reputation for honey-making, the uplands behind the village being clothed with heather that seems to suit the bees. (34, 61, 93.)

Cove and Kilcreggan (862) together form a police burgh. The latter village fringes the tip of the peninsula between the Gare Loch and Loch Long, while Cove lies about a mile up Loch Long. Kilcreggan is one of the most popular watering-places on the Firth of Clyde. Its lovely situation and the ease with which it can be reached from Glasgow are two of its greatest attractions.

In less than an hour after leaving the smoke-laden atmosphere of Glasgow one can step on to the pier at Kilcreggan. The village is therefore really a suburb of Glasgow, for many men with businesses in the city have their homes in Kilcreggan. The shore between the village and Cove is lined with fine villas and mansions. (13, 34, 62, 63, 134.)

Cumbernauld (1214) is one of the oldest towns in Dumbarton. It is situated on the main road from Glasgow to Falkirk, Stirling, and Edinburgh. It was made a burgh of barony so long ago as 1649. Near Cumbernauld a remnant of the ancient Caledonian forest remained until comparatively recent times. The district used to be noted for the savage, white cattle of which a few specimens are still to be seen at Bothwell. In 1571 a writer complains of the keeping of the "quhit ky and bullis, to the gryt destructione of polecie and hinder of the commonweil. For that kind of ky and bullis hes bein keipit this money yeiris in the said forest." Cumbernauld used to be noted as a hand-loom weaving centre, but nowadays quarrying and mining are the chief industries. (26.)

Dumbarton (21,989) until recent years was the only important town in the shire. In early times it was the greatest town in the Clyde valley till Glasgow asserted its supremacy chiefly through its cathedral and its university. In 1222 Dumbarton was made a royal burgh by Alexander II, and received other charters from succeeding kings. Its site is a flat, alluvial plain near the confluence of the Leven and the Clyde. The town is not a particularly well-built one, for its main end in view has generally been connected with trade. Shipbuilding is the principal industry, the two great firms being Denny and McMillan. General engineering is also carried on in the town, and there are other trades connected with the sea, such as sail-making, boat-building, and paint-making. From the Castle Rock a long pier juts out into the navigable channel of the Clyde. There are also a dock and a good quay. Dumbarton Academy is one of the best known secondary schools in the west of Scotland. (15, 44, 45, 55, 57, 76, 82, 91, 93, 96, 101, 102, 103, 106, 107, 108, 119, 128, 129, 130, 133, 134, 136, 138.)

Duntocher (3092) lies at the foot of the Kilpatrick Hills, two miles east of Old Kilpatrick. Although the little town itself is not attractive, the surroundings are exceedingly pretty. The place must have been an important station in the days of the Roman occupation, for many Roman relics have been found in the vicinity of the village. A very old bridge in the village is also called a Roman bridge, but the evidence is not convincing. It is, however, believed by some antiquaries to be perhaps the oldest bridge in Scotland. The town was one of the earliest seats of the cotton industry in Scotland, and by the middle of the nineteenth century had four large mills within a mile of each other. At the present time the inhabitants of Duntocher are engaged in a great variety of industries. These include coal-mining, quarrying, lime-working, engineering and shipbuilding, for the town is near the great yards and factories of Clydebank. (101, 113.)

Garelochhead (860) is a pretty village at the head of the Gare Loch. In summer time it is a popular watering-place, the climate being extremely mild. (76, 81.)

Helensburgh (8529) is situated at the mouth of the Gare Loch exactly opposite Greenock. The town arose towards the end of the eighteenth century on land belonging to Sir James Colquhoun, after whose wife the town took its name. It was intended originally to create an industrial town, and it was announced that "bonnet-makers, stocking, linen, and woollen weavers" would "meet with proper encouragement." But the town never attained any success in manufactures. Its position, mild climate, and beautiful surroundings have been its greatest assets, and consequently Helensburgh is to-day almost entirely residential. It is perhaps the best planned town in the west of Scotland. Long streets run parallel to the shore and are intersected at right angles by shorter streets.   The thoroughfares are broad, well-kept, and in many cases planted with trees.   On the front facing the sea, stands a fine granite obelisk in memory of Henry Bell, provost of the burgh, 1807-1809.   (4, 34, 57, 76, 82, 124, 128, 134, 136, 138.)

Helensburgh from the south-east

Jamestown is a small town in the Vale of Leven, on the left bank of the river, and a mile north of Bonhill.   Its rise has been due to the dyeing and calico-printing industries of the district.   (96.)

Kirkintilloch (11,932), the name of which is a corruption of *caer-pen-tulach*, "the fort at the end of the ridge," is one of the oldest in the country.   It has been a place of some importance from the time of the Romans.   Antonine's Wall ran through the site of the town.   As early as 1170 Kirkintilloch was made a burgh of barony by William the Lyon.   William de Comyn, Baron of Lenzie, was the most powerful nobleman of the district at this time.   At the beginning of the fourteenth century the

10—2

district passed into the hands of the powerful Fleming family, Earls of Wigtown. In 1745 Kirkintilloch suffered at the hands of the Highlanders. The inhabitants find occupation in engineering works, saw-mills, chemical works, and muslin mills. (82, 113, 130, 131, 134, 139, 140.)

Kirkintilloch

Lenzie is a village partly in Dumbarton and partly in Lanark. It consists largely of villas and cottages, and it may be considered mainly as a residential suburb of Glasgow. In Lenzie and the neighbourhood are several large public institutions, the finest of which is Woodielee Lunatic Asylum, a very handsome building situated in beautiful grounds. (127, 136.)

Luss (553) has one of the most charming situations in the west of Scotland. It lies at the foot of Glen Luss, and has an outlook over some of the prettiest islands of Loch Lomond. In the background are several mountains over 2000 feet in height. Luss and the vicinity suffered severely from the expedition sent by Haco across the Tarbet isthmus to bring fire and sword on Loch Lomondside. Coleridge, Wordsworth, and his sister Dorothy spent a night at Luss Inn in 1803. Queen Victoria also passed a little time at Luss on her tour from Inverary to Balloch in 1875. The village is a favourite resort of anglers and tourists. A short distance inland a band of slate outcrops, and is worked for roofing slates. (34, 49, 53, 118, 126.)

The Pierhead at Row, Gare Loch

Milngavie (4530) was in Stirlingshire until 1891, when the Boundary Commissioners altered the county boundary so as to include the whole of New Kilpatrick parish. The town is seven miles north-north-west of Glasgow, and stands nearly 200 feet above sea-level on the Allander Water, a picturesque tributary of the Kelvin. The surrounding country is open and attractive, with fine views of the Kilpatricks, the Campsies, and the Highlands. The town has therefore developed rapidly in recent years as a residential suburb of Glasgow. In the period between 1901 and 1911 the population increased by 23 per cent., one of the largest burgh increases in Scotland. In the town or in the immediate vicinity some dyeing, bleaching, and distilling are carried on. Less than a mile to the north are the two great reservoirs that store for Glasgow the waters of Loch Katrine. (3, 11, 26, 129, 134.)

The Hydropathic, Shandon

Milton is a village in Old Kilpatrick parish between Dumbarton and Bowling. The village claims recognition in the history of industry as being the first place in Scotland to possess a power-loom. On the slopes of the Kilpatricks behind Milton there are fine examples of the mysterious "cup and ring" markings.

Old Kilpatrick village (1916), situated four miles south-east of Dumbarton, is said to have been the birthplace of St Patrick.

In 1679, it was made a burgh of barony, but it allowed its privileges of town government to lapse. It is now a residential village largely for the better class of workers in Clydebank and other large industrial centres. (107, 118.)

Renton (4903), in the Vale of Leven between Dumbarton and Alexandria, was founded in 1782 by Mrs Smollett of Bonhill, and named in honour of her daughter-in-law, Miss Renton of Lammerton. Like the other towns of the Vale of Leven, Renton is engaged in calico-printing, bleaching, and dyeing. The town is not specially attractive, but the surroundings are exceedingly pretty. (82, 94, 138.)

Rosneath (724) is situated near the mouth of the Gare Loch and on the western shore. From some points of view it is the prettiest village on the loch. There is a lovely bay just south of the village, and the woods in the vicinity are particularly beautiful. A mile from the village stands Rosneath Castle, one of the principal seats of the Duke of Argyll. (53, 61, 76, 93, 107.)

Row (1193), an entirely residential village, rivals Rosneath in its charming surroundings. The bay affords a safe anchorage for yachts. (34, 58, 81, 105.)

Shandon, from the Gaelic *sean dun*, "old fortress," is a village on the eastern shore of the Gare Loch. North and south of Shandon the shores of the loch are dotted with large mansions, the summer homes of wealthy merchants. One of the finest was the seat of Robert Napier, the famous engineer. On his death his art collections were sold, and the house converted into Shandon Hydropathic, one of the best known health resorts in the west of Scotland.

Yoker is on the borders of Dumbarton and Renfrew. The town is continuous with Clydebank, of which it really forms a part. (10, 82.)

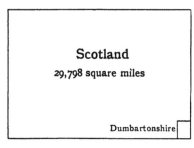

Fig. 1. Comparative areas of Scotland and Dumbartonshire

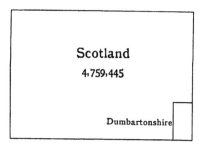

Fig. 2. Comparative Populations of Scotland and Dumbartonshire

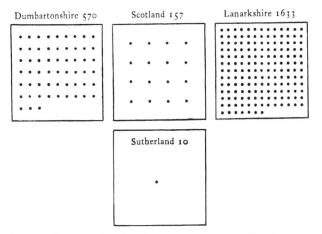

Fig. 3. Comparative density of population in Dumbarton-shire, Scotland, Lanarkshire and Sutherland.

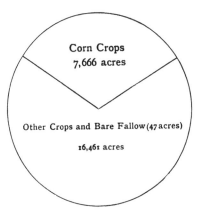

Fig. 4. Proportionate areas of corn and other crops in Dumbartonshire

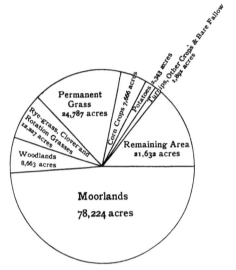

Fig. 5. Proportionate areas of principal kinds of
vegetation in Dumbartonshire

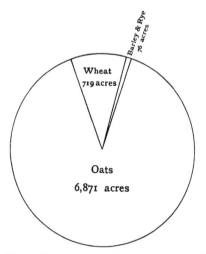

Fig. 6. Proportionate areas of oats, wheat, and barley
in Dumbartonshire

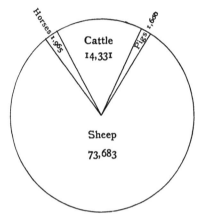

Fig. 7. Proportionate numbers of sheep, cattle, horses, and pigs in Dumbartonshire

www.ingramcontent.com/pod-product-compliance
Ingram Content Group UK Ltd.
Pitfield, Milton Keynes, MK11 3LW, UK
UKHW042145280225
455719UK00001B/111

9 781107 678774